SpringerBriefs in Energy

SpringerBriefs in Energy presents concise summaries of cutting-edge research and practical applications in all aspects of Energy. Featuring compact volumes of 50 to 125 pages, the series covers a range of content from professional to academic. Typical topics might include:

- A snapshot of a hot or emerging topic
- A contextual literature review
- A timely report of state-of-the art analytical techniques
- An in-depth case study
- A presentation of core concepts that students must understand in order to make independent contributions.

Briefs allow authors to present their ideas and readers to absorb them with minimal time investment.

Briefs will be published as part of Springer's eBook collection, with millions of users worldwide. In addition, Briefs will be available for individual print and electronic purchase. Briefs are characterized by fast, global electronic dissemination, standard publishing contracts, easy-to-use manuscript preparation and formatting guidelines, and expedited production schedules. We aim for publication 8-12 weeks after acceptance.

Both solicited and unsolicited manuscripts are considered for publication in this series. Briefs can also arise from the scale up of a planned chapter. Instead of simply contributing to an edited volume, the author gets an authored book with the space necessary to provide more data, fundamentals and background on the subject, methodology, future outlook, etc.

SpringerBriefs in Energy contains a distinct subseries focusing on Energy Analysis and edited by Charles Hall, State University of New York. Books for this subseries will emphasize quantitative accounting of energy use and availability, including the potential and limitations of new technologies in terms of energy returned on energy invested.

More information about this series at http://www.springer.com/series/8903

David C. Green

Home Energy Information

Measuring and Managing Energy Consumption in Residential Buildings

Springer

David C. Green
Green Management Services, Inc.
Fort Myers
Florida
USA

ISSN 2191-5520 ISSN 2191-5539 (electronic)
ISBN 978-3-319-11348-7 ISBN 978-3-319-11349-4 (eBook)
DOI 10.1007/978-3-319-11349-4
Springer Cham Heidelberg New York Dordrecht London

Library of Congress Control Number: 2014951915

Printed on acid-free paper

Springer is part of Springer Science+Business Media (www.springer.com)

Preface

What's your *number*? Is there a *number* that can tell you how well you conserve energy, how energy efficient your home is? Yes, there is. This book will help you define that number and continuously improve it. Having spent several years building web-based Energy Information Systems for large organizations I found many ways to look at the data produced by the energy we consume. Collecting, analyzing and reporting energy consumption data and turning that into valuable information is the key to conservation. The majority of our energy consumption takes place in the workplace and has to do with business processes around the world. The energy information generated by those processes is complex and produces massive volumes of data. Conservation efforts in the workplace often get "derailed" by budget cuts or lack of expertise. This is why I feel the best place to start working on energy conservation is in the home. So I wrote this book to describe a simple and easy way for any homeowner or renter to effectively collect and analyze their energy consumption. Once they have done so and taken steps to conserve they can reapply the same methods to observe the results of their conservation actions. Teachers and students at almost any level may also find it useful for exciting classroom projects spanning the entire school year. The concepts described in this book are so simple, easy to understand and relevant to everyone that there is a perfect opportunity for them to proliferate across the globe through social media outlets. My hope is that these methods, once proven in the home, will provide valuable insight and inspiration for responsible parties in the workplace to take action toward energy conservation in an easy and effective way. The book begins with an introduction to the entire process of collecting energy information, analyzing it, making changes to your consumption and utilizing renewable energy sources. There is a chapter that explains in detail the units of measure used to track energy consumption and a chapter that shows how to analyze that data and turn it into useful information. Next I show how to use various devices to focus on your energy consumption in detail. This allows you to find the trouble spots and where you might be able to improve.There is a chapter explaining various methods used to conserve energy in the home and a chapter on renewable energy sources and how they can be used to help conserve. Finally a conclusion pulls all this together and explains how social media can help "spread the word" about how easy it can be to conserve energy.

Green Management Services, Inc., Fort Myers, FL, USA
dcgreen@dcgreen.com
David C. Green

Contents

Chapter 1
Introduction

Abstract Home energy information is information collected or calculated in the course of analyzing our home energy consumption. We use energy information in a cyclical fashion in that it is collected and analyzed over and over again. Utility companies measure energy consumption in standard units of measure depending on the type of energy consumed. We can analyze those values and others derived from them over time to track the results of conservation and renewable energy efforts.

1.1 Home Energy Information Cycle

The Home Energy Information Cycle is a recursive method of monitoring, measuring and taking action. The basis for this cycle is the energy information itself. It's the "measuring stick" for deciding how well we are doing when it comes to energy conservation (Fig. 1.1).

The Home Energy Information Cycle begins with the utility company sending us a bill for services. Most of us monitor our energy consumption using the monthly invoices received from the companies that sell us the fuel to provide the energy to our homes. If the invoice is from an electric company then it will also state the kilowatt-hours (kWh) of electricity used during the month. If it is from a natural gas company it will also show the number of therms (or some other unit of measure) of gas consumed during the month. A propane or fuel oil supplier would send us an invoice for the number of gallons of propane delivered. We can also monitor our energy consumption by observing changes to our electric and natural gas meters. Gauges on our propane and fuel oil tanks tell us how much we have consumed. This is the information we need to start analyzing our consumption.

We analyze the information in the bill pertaining to the amount of energy consumed because it pertains to a particular month of the year and a particular location. Varying temperatures from month to month can make a dramatic difference in the amount of energy consumed. Also, our homes consume more or less energy depending on their size. Energy managers for large organizations typically "normalize" the energy consumption values with respect to temperature and the amount of area serviced measured in square feet (SF). In this way, we can compare the rate of consumption month to month and location to location. This allows us to establish

D. C. Green, *Home Energy Information,* SpringerBriefs in Energy,
DOI 10.1007/978-3-319-11349-4_1

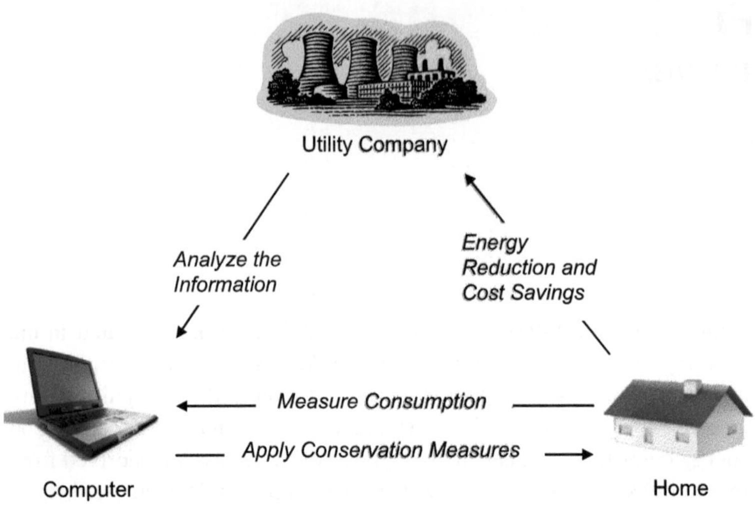

Fig. 1.1 Home energy information cycle

a baseline from which to measure improvement as we make changes to particular sources of energy consumption.

Measuring the energy consumption directly from its source gives us the information we need to decide when and if changes are required. Home owners can use small, inexpensive power meters to measure the electric power consumption of any appliance that plugs in to a wall outlet. The devices are safe for anyone to use and measure the electricity consumed in kWh. This makes it easy to compare the differences in electric consumption for all those plug-in appliances in the home. Devices are also available to measure the hot or cold air leaking into the home around windows and doors. This is an excellent way to focus in on when to take action and what actions to take in order to conserve energy.

Taking actions to try to conserve energy is a popular subject and has been for several years. There are many books available about how to make your home more energy efficient. There are even books about how to live completely "off the grid" and consume absolutely no power … at least not from power companies. Insulating is one of the most effective ways of improving your home's energy efficiency. If you own an older home your insulation is probably not as effective as the insulation available today. One inexpensive way to start conserving is to replace older incandescent light bulbs with CFLs or LED light bulbs. These light bulbs not only use less electricity but they last much longer as well. Changing out an older refrigerator for a newer, more energy efficient model will likely make a measurable difference in the amount of energy your home consumes.

After taking actions to conserve it is important to circle back and re-measure energy consumption using the devices mentioned or any other methods to verify that changes actually resulted in improvements. Small changes spread over several appliances will add up to significant improvements. Diligence is important since certain improvements are applicable only at certain times of the year. The process should be continuous month after month throughout the year.

If we dedicate ourselves to the Home Energy Information Cycle we repeatedly monitor energy information from our energy providers, take actions to conserve and measure consumption rates. This process should yield a noticeable decrease in the amount of energy consumed over time.

1.2 Home Energy Units of Measure

We express Home Energy Information in units of measure. Energy providers measure consumption of electric and fossil fuels such as gas, propane and fuel oil in standardized units so that we all understand exactly what we are paying for. We use degree days (DD) to understand the effect of hot or cold outdoor temperatures. We use living area measured in square feet (SF) to understand how much space we are heating or cooling in a particular location. Calculated units of measure tell us even more about our energy consumption by allowing us to compare our energy efficiency over time or from one location to another. Understanding these units of measure is the key to conserving energy at home.

Electricity consumption is probably the most difficult to understand of all units of measure. Providers measure electricity consumption in *kilowatt-hours* (kWh). A 100-watt light bulb consumes one kilowatt-hour of electricity in 10 h [1]. Most providers price our electricity in terms of kilowatt-hours. A television consumes approximately 100 W of electricity over the period of one hour or 0.1 kWh.

A *British thermal unit* (BTU) measures heat. The most common units used to measure natural gas consumption by utility companies are *therms* (1000 BTUs). Decatherms (dkt), equal to 1,000,000 BTUs, is also used. It is sometimes referred to as MMBTUs which stands for one thousand thousand BTUs or 1,000,000 BTUs [1]. The BTU is a common unit of measure. We can convert the consumption of any energy source in any units to BTUs.

Natural gas meters actually measure the volume of gas consumption in *cubic feet*. Some utility companies express this consumption as 100s of cubic feet (CCF). One hundred cubic feet is 1 CCF. 100 CCF is equal to 1 therm or 1,000 BTUs [1].

Providers price and sell propane and fuel oil by the *gallon*. These fuels are a little tricky to monitor effectively since providers do not typically deliver or invoice on a monthly basis. Their delivery schedule is usually dependant on how much fuel the customer consumes since the holding tanks limit the amount of fuel the customer can keep on-hand. Energy managers use conversion factors to express gallons of these fuels in terms of the number of BTUs given off when burned [1].

We use *degree days* (DD) to relate the outside temperature to the energy required to heat or cool dwellings. Obviously, the colder it is outside the more energy is required to heat a structure and the hotter it is the more energy is required to cool a structure. Degree days (DD) are a measure of how many degrees and how many days the outside air temperature was below or above a certain base temperature, usually 65°. If the mean daily temperature is lower than the base temperature then the degree days (DD) are called "heating degree days". If the mean daily temperature is above the base temperature than the degree days are called "cooling degree days". The Mean Daily Temperature is the average of the highest and lowest temperatures in a 24-hour period [2].

We measure the area of a dwelling in *square feet* (SF). Service area refers to the number of square feet of a dwelling serviced by a particular energy source, electricity, natural gas, etc. Some areas may not be heated or cooled so should not be used to calculate the service area. The service area for electricity may differ from the service area for natural gas. We use the service area to relate the size of a structure to the energy it consumes.

We can use these units of measure along with some formulas to accurately turn detailed energy consumption data into valuable energy information. Energy information derived from these measurements allows us to make qualified decisions regarding our energy conservation efforts.

1.3 Measuring Energy Consumption

We measure energy consumption in order to turn that data into valuable home energy information used to help us conserve. Our measurements pertain to energy consumption that occurs over any number of timeframes. The most common timeframe being the "month" since we receive invoices from electric and natural gas providers every month. As energy conscious homeowners, we collect and record these monthly values in order to analyze the energy efficiency of our homes.

We need to keep track of specific values on the invoices issued by electric providers each month. There is a lot of information on an electric company invoice. However, the only values we are interested in are the start and end service dates along with the actual kWh consumed for the month. Service dates or "billing dates" for these invoices rarely start exactly on the first day of each month and end exactly on the last day.

There are also similar specific values on the invoices issued by natural gas providers each month. They are the start and end dates of service and the therms or CCFs consumed during that month.

Collecting monthly consumption values for propane and fuel oil is a little more difficult, in some respects at least. Propane and fuel oil providers usually issue invoices after each delivery. That may be once or twice a month or once every two to three months. So in order to get a monthly consumption measurement we have to just simply look at the gauges each month on the first of the month to determine how much we have consumed. The good thing is, this eliminates the need to track the different start and end service dates for each month.

We collect degree days (DD) from the web site http://degreedays.net. This site measures the degree days (DD) for weather stations in any area and makes them available in convenient reports that we can download. So, we can always get a reasonably accurate measurement of how temperature relates to our energy consumption data.

We can measure the area of our homes by using the property tax information data or by simply measuring the area of each room in square feet (SF) then adding them all together. Keeping in mind that the service area for electric consumption should not include areas where there is no electric service or electric heating and cooling. The same principle applies to all service areas for all sources of energy.

We collect these measured values and record them in a spreadsheet or some form of tabular record. We record the measurements for each month. Computer savvy homeowners may wish to create a database to hold these values. We will combine these measurements into formulas for analyzing our energy consumption over time and from place to place. We will be able to compare our energy efficiency month to month for our location and with those in other locations following the same method of measuring and analysis.

1.4 Analysis

The purpose of analyzing energy consumption measurements is to turn that data into energy information that can help us make better decisions about how to conserve. If we can determine how energy efficient our home is for any particular month and compare that to previous months or even to other homes, then we have a method of monitoring our conservation efforts. We can then establish our own *baseline* to compare against over time. We can compare our energy efficiency to other homes anywhere in the world and use that information to inspire us to conserve even more. In order to make this comparison I've found we can express our home's energy efficiency as a *home energy index* (HEI). This is the *number* that we must continuously try to improve (lower).

Creating the HEI requires all of the units of measure we discussed previously that apply to your home. Most everyone has electricity so you need to have recorded the kWh used each month. If you use natural gas for heating or hot water then you must record the therms used each month. You will also need to know the total degree days (DD) for all days of the billing cycle for these energy sources. Also, make sure you have a record of the area in square feet (SF) being serviced by each energy source. We express the home energy index (HEI) as BTUs per SF DD.

In order to calculate the energy index (EI) for each energy source we must convert the energy consumption values into BTUs. You can do this easily at the web site http://www.onlineconversion.com/energy.htm. To convert electric units, 1 kWh = 3,412.141 BTUs. Also, to convert natural gas units, 1 therm = 100,000 BTUs. Similar conversion formulas are available for propane and fuel oil. So, we can calculate an EI for any energy source using formulas similar to those shown below.

$$\text{Electric EI} = \left(\text{kWh} \times 3{,}412.141\right) / \left(\text{SF} \times \text{DD}\right)$$

$$\text{Natural Gas EI} = (\text{therms} \times 100{,}000)/(\text{SF} \times \text{DD})$$

By using these formulas we are actually calculating or "measuring" the number of BTUs consumed for each square foot of area and each degree day for each month for each energy source. In doing so, we remove or *normalize* the variations in home size and temperature from our analysis of energy consumption. We can now compare our consumption over time no matter what temperature change occurred month to month. We can also compare energy consumption from home to home no matter what the difference in area. Since we converted the energy units to BTUs we can add these energy source EIs together to get an HEI for the entire home. This is the number that is important. It's the number we strive to *lower* each month. Each month that our HEI is lower than the previous month we are on the right path to conserving energy.

1.5 Conserving Energy

Any homeowner can take steps to lower their HEI, conserve energy and help protect our environment. Coal provides 40 % of the worlds electricity and 44 % of the carbon dioxide (CO_2) emitted by fossil fuels comes from coal, the cheapest and dirtiest fuel [3]. CO_2 is the chemical byproduct of carbon oxidation (burning). Coal is nearly pure carbon. After we learn how to analyze our energy consumption we can start conserving, observing the results and feel good about our efforts to reduce these dangerous emissions. Evaluating our energy consumption throughout the year leads to clues about where we need to *seal*, *insulate* and *replace* items. A good place to start looking is in the attics, basements or crawlspaces. We can also make improvements in the yard and with our windows and doors. Every little bit helps and for each small improvement we lower our energy costs as well. Annual energy cost in the U.S. for a typical home is $ 2,200. Half of that is heating and cooling [4]. Sealing leaks in the home is the best way to start saving on our heating and cooling bills.

A good portion of the heating and cooling losses in the home are due to air leaks. Leaks occur around windows, doors, wiring, siding, chimneys, foundations and any place that is a barrier to the outside. One of the best ways to find them is to just walk around your home looking for cracks in the foundation or gaps around windows and doors. Sealing these up from the outside with caulking will stop the "bad" outside air from entering your home and the "good" inside air from escaping. On the inside, look for worn or missing weather stripping and gaps around pipes or wires leading under the house. You can find these leaks from the inside using a thermal leak detector as well. During the times of the year when temperatures are extreme just point the detector around the edges of windows and doors and look for changes in temperature readings. This is also a good way to find areas of the home that need more insulation.

Insulating is the next best way to conserve energy and save on energy costs. Insulation comes in various types and some are dangerous to work with. Safety is the most important aspect of working with insulation. Older homes just do not have the insulation recommended by today's standards. Insulation may also be deteriorated or ineffective for any number of reasons. It's important to inspect your home's insulation starting in the attic. The top of the house is where all the heat goes in the winter and where all the heat comes from in the summer. So, take steps to insulate your home in such a way to prevent heated air from entering the attic. Also insulate to prevent air in the attic, heated by the suns rays, from entering the living area of the home. We should look for opportunities to insulate forced air ducts and hot water pipes to make them more efficient. At the bottom of the house in your basement or crawlspace cold damp air can creep up through the floor and make your living space uncomfortable when it is heated. Since cool air tends to fall downward cool comfortable air can escape through the floor. Air movement through the home due to hot air rising and cold air falling can be used to our advantage in ventilating our homes.

Ventilation is an important aspect of the overall health of our home. It prevents dangerous airborne particles from building up inside the living area as well as provides heating and cooling advantages during moderate seasons. The *chimney effect* exists when warm air rises out of the home through the attic vents and pulls cool air up into the home from windows, floor vents or basement doors [5]. Whole house fans accelerate this effect to force air out of the attic vents faster. It stands to reason that the same effect could work in reverse during colder months to draw air warmed by the sun shining on the roof out of the attic down into the living area. Some of these ventilation techniques may require thermostats and switches in order to be effective.

Electronic devices play an important role in conserving energy. We find them throughout the home as programmable thermostats, motion sensors, photocells, timers and others. These devices turn appliances on or off automatically at the appropriate times to save energy. Appliances are not the leading consumers of electricity but they account for a considerable amount. It's a good idea to inventory all your appliances and evaluate which ones need special attention, regular maintenance or replacing. You can even estimate the electricity used by these appliances if they have a label showing the wattage of the appliance. It is especially important to maintain the refrigerator by keeping the coils clean. A refrigerator runs more efficiently if the freezer is full. Clothes dryers are huge consumers of electricity and work much more efficiently when the dryer vent is clean. Replace old incandescent light bulbs with more efficient ones that last longer. Outside the home, cleaning the air conditioning coils helps it run more efficiently.

Making changes outside the home can help conserve energy. You can plant trees and shrubs in critical places to block wind or shade the house. Awnings over windows that face the sun help to reduce the solar rays entering the home in the summer. Special windows and screens are available to help prevent unnecessary solar heating. On the other hand, solar heating can be an effective way to supplement heat in the winter months. Keep south facing windows clear of trees and shrubs to allow sunlight into the living areas. If you are the technical type of person you may wish to experiment with solar panels or solar heating arrays to supplement your electricity and hot water supply. Use monitoring devices to track the effectiveness of your experiments.

Monitoring devices are a great way to track the progress of all the energy conservation improvements you make to your home. Small handheld devices can monitor the electricity used by plug-in appliances. You can use the same thermal detector mentioned earlier to measure the temperature difference created by your heating or cooling system. Whole house monitors are available to track energy consumption in real-time. Some of the more sophisticated devices provide reports and even online access to the data. These devices, some of them quite inexpensive, are an important aspect of conserving energy because they provide instant feedback on how well our conservation efforts are doing. This instant feedback will provide an early clue as to whether or not we are lowering our HEI.

If our energy conservation efforts successfully lower our HEI month to month then we are on our way to protecting the environment and saving money. We should think of this as a continuous process rather than a quick fix. Reducing the amount of electricity used to heat or cool our homes dramatically reduces the amount of coal used and thus the amount of CO_2 emissions produced. Obviously, the methods we use during this process are available to everyone and if everyone takes advantage of them our planet will benefit greatly. So, passing on what you learn about energy information, conservation and monitoring devices to your friends and family is extremely valuable.

1.6 Monitoring Devices

Any homeowner can use devices to monitor energy consumption in real time throughout the home. Access to this information is a great way to check the results of conservation measures homeowners have taken. Of course, the most obvious monitoring devices are the meters and gauges themselves. These devices installed by the utility service companies can provide valuable information to us. It's easy to step aside and read the values on your electric meter, record them, and then come back at a later time record the values again and subtract the two to find out how much electricity the home used during that time. The same thing is true for natural gas meters. However, propane and fuel oil might be a little more difficult to measure. Some utility companies provide web pages that let the homeowners monitor their utility consumption day by day or even hour by hour. Also there are a number of whole house monitoring devices that will monitor the electric consumption for the whole home. On a smaller scale, there are devices we can use to measure the consumption of individual appliances or circuits in the home. The important thing to remember is that a homeowner can be proactive about monitoring energy consumption by taking as little effort as watching the electric meter.

Electric meters, usually mounted outside the home, are an easy and convenient way to track electric consumption in real time. For instance, if you want to see how much electricity your home is consuming in a particular hour you can simply step outside and record the numbers on the electric meter then wait one hour and step outside to the record numbers again. Subtract the two numbers and that will give you the amount of kilowatt hours or watts consumed during that hour. The same thing goes for the natural gas meter. If you have a gas stove in your kitchen and you want to see how much natural gas the stove consumes you can check the natural gas meter reading when no other gas appliances or systems are running. Then turn the stove on for 10 to 15 min and then check the meter again. Subtracting values will tell you how much gas the stove consumed during that time. I use propane to fuel my gas fireplace. I have a small 100 gallon propane tank which has a gauge attached so that I can check when to order more propane. I've also used this to measure how much propane the fireplace uses. I step outside and check the value on the gauge then turn the fireplace on. I usually burn the fireplace for least a few hours in the evening during the winter. After a period of time, 1 to 2 hours, I will step outside and check the gauge again to see how much propane I used during that time. That tells me the rate of consumption of propane that my fireplace is using. Fuel oil consumption might be somewhat more difficult to measure. I can remember many years ago when I had a fuel oil furnace and I had a 250 gallon tank. The way that I checked the amount of oil in the tank was I dropped a measuring stick down into the top of the tank and read the markings on the stick. Aside from these primitive methods of measuring consumption, these are the most convenient types of monitoring devices available to us because almost every home has them. This is how the utility companies track our consumption and decide how much to charge us. Take advantage of these opportunities to monitor your energy consumption and use the information to help you conserve.

Some utility companies will help you monitor your energy consumption for you. My electric company provides web pages that show my electric consumption on a weekly daily or even an hourly basis. This is an excellent tool to see how much electricity I'm consuming throughout different parts of the day. Smart electric meters allow this to happen. Smart meters provide feedback to the utility companies about electric consumption. There are also whole house energy monitors available that plug into your electric meter or into your circuit panel. These devices will monitor the energy consumption of your home as a whole. Some of these devices are WIFI enabled so that you can view the consumption in real-time on display devices conveniently located in the home.

Some devices called data loggers can plug into your circuit panel and monitor the whole house electric consumption or the consumption associated with any particular circuit. On a smaller scale, you can use devices to measure the consumption of individual appliances throughout the home. Small devices are available that plug into the wall and then allow you to attach a plug-in appliance to them. The device will record the kilowatt hours consumed and the amount of time you monitored the appliance. Also, you can use thermal leak detectors to monitor how much hot or cold air is coming out of a forced air vent. If you know that your forced air ducts are leaking, take these thermal measurements prior to sealing and insulating your ducts. Then take the same measurements after you make the repairs. You should see an improvement in the amount of heat transferred to your living area. More sophisticated devices are available to monitor any number of appliances throughout the home and send the information by WiFi signals to a central collection station and display for the homeowner to observe.

So you can see there are a number of options available for the homeowner to monitor energy consumption and collect the information required to measure the effectiveness of the conservation steps they have taken. This information provides feedback on the efforts made to seal and insulate your home or replace high energy consuming appliances. Homeowners should inventory all the energy consuming appliances in the home and record small time frame consumption values for these appliances so that they compare them to see which appliances are high consuming appliances. Whole house monitoring devices help us understand how high energy consuming systems such as heating and air conditioning effect our energy consumption at different times during the day. For instance, it might be interesting to compare the electric consumption for one hour during the hottest part of the afternoon on a hot summer day to the electric consumption for one hour during the middle of night. These comparisons help homeowners understand more about energy consumption in the home and what they can do to lower costs.

1.7 Renewable Energy Sources

Using renewable energy sources is an excellent way for the homeowner to lower energy costs. However, it takes a more technically inclined homeowner to take advantage of these renewable energy sources. Installing renewable energy sources on and around your home can supplement your energy supply. It is only possible when they are affordable and available. The most popular renewable energy source is solar power. Solar panels mounted on the ground or roof can supplement the electric supply for the home. In some cases solar systems can also supply energy back to the electric grid and earn credits on the homeowner's electric bill. Other, more specific, systems like solar water heaters, solar water pumps and solar attic fans also take advantage of the sun's rays. Wind power systems mounted near the home can provide electric power in a similar way to solar panels. Certain circumstances may allow the homeowner to take advantage of hydroelectric power. With any of these electrical supply systems it's important to have storage batteries in order to save the power during times that the sun, wind or water is not available. Once your renewable energy systems are in place, it will be helpful to utilize data management systems to track the energy produced. Also you will need controls and switches to manipulate the energy in such a way to make it most beneficial to you. Many renewable energy systems have backup generator power to supplement the renewable energy source when needed. Probably the oldest example of renewable fuel is wood. Wood is a renewable resource that people have used for thousands of years as a heat source. In any case utilizing renewable energy requires the homeowner to have some technical knowledge and the willingness to experiment. I suggest starting with a simple solar power system to understand the principles of creating electrical energy from the sun and storing the energy for later use.

Using solar power to supplement electricity needs involves many different criteria. One of the most important, if you intend to put the solar panels on your roof, is to determine the orientation of your roof to the sun. The most ideal situation is to have a roof slope that faces south. This allows you to get the most hours of sunlight per day onto your solar panels. You should also investigate the incentives from the federal, state and local governments. These incentives sometimes allow tax credits for installing solar panels at your home in order to offset the cost of the equipment. There are different types of solar power systems. *Off grid* systems always include backup batteries to store the energy for use when the sun is not shining. *Grid tied* systems do not have battery backups but sometimes provide energy back onto the electric grid. When the sun is shining, this can result in earning credits on your electric bill. In *grid connected* systems there is a connection to the utility grid but also a bank of backup batteries. There are also portable solar systems that can charge small electrical appliances like phones computers radios and televisions. A simplified solar power system includes the solar panels themselves, a charge controller, an inverter to convert the solar panel direct current electricity into alternating current that is useful for the home, batteries to store the electricity and the wiring to connect them all. Some systems, more simplified and specific, do not need the inverter or the batteries.

Much smaller, more specialized, solar power systems help to provide energy to specific areas of the home. For instance, a solar water heater consisting of a series of black heat conducting tubes placed on the roof circulates liquid through a heat exchanger transferring heat from the sun's rays into water. These systems are useful for supplementing the home hot water supply or heating a swimming pool. Solar water pumps are a great way to pump water for irrigation systems. Another excellent small solar power system is the solar attic fan to add a little extra power to the process of purging hot air from the attic. The possibilities for using solar power are almost limitless. There are even examples of solar electric fences to place around gardens and prevent wild animals from destroying garden plants. Another more passive example of solar energy use is the solar light tube which is similar to a skylight. Light travels down a tube and then into a home and acts like a recessed ceiling light.

Wind power is another excellent way to supplement the home electrical energy needs. Wind systems, tied to the electric grid just like solar power systems, also provide energy back to the electric grid and earn credits on the homeowner's electric bill. It's important to determine the wind speed in your area to decide if you are in a location where utilizing wind as an energy source is beneficial. Obviously, if your home is high on a hill the more wind you will have. Similar incentives exist for these systems as with solar power systems. Wind generating systems are usually mounted on tall towers 24 to 50 ft tall adjacent to the home.

If you happen to have a fast flowing stream next to your home it may be possible to use hydroelectric power to supplement your home energy needs. Criteria such as the head, flow, distance and location are all things to consider in designing your hydropower system. There needs to be a considerable elevation drop and sufficient flow of water in order for the system to turn a generator and produce electricity. But these systems are available.

Many solar, wind and hydro power systems require storage batteries to store electricity in times when the sun, wind and water are not available. Deciding what kind of batteries, how to connect them and how to care for them is important part of building the systems. The systems usually require large battery banks in order to store enough energy to make it useful in times of need. You need to be able to monitor the voltage and the state of charge of the batteries at all times.

Management systems are available to monitor the state of your renewable energy system. These small energy management systems can track the power in and out of your home and the status of your storage batteries. These management systems along with the controls and switches that make the whole system work together are an important part of utilizing renewable energy to supplement your home energy needs. They provide early warning for situations where you might run out of power.

No renewable energy system would be complete without a backup power source. This is usually an electric generator fueled by gasoline or diesel. An even more efficient way to fuel the generator would be to create your own biodiesel from plant waste or to distill your own alcohol. I have an apple tree that produces way more apples than I could possibly eat in a given season. Most of the apples fall to the ground and rot. I've often wondered if I could distill those apples into pure alcohol to burn in the generator to produce backup power. I suppose there are more convenient ways to provide backup power.

One of the most convenient forms of fuel is wood. Homes located in a wooded area have a ready abundance of wood waste to burn for fuel. This is perhaps the oldest renewable fuel known to man. For many years people heated homes primarily by burning wood. Now more modern wood burning fireplaces outside the home can transfer the heat through pipes into a home's heat exchanger and directly benefit the heating system. Wood is an excellent source of heat energy. However some critics claim that the carbon dioxide emissions produced by burning wood offset the benefits with regard to protecting the environment.

So you can see that there are many options when it comes to utilizing renewable energy to supplement our home's energy needs. It takes a technical mind to understand and develop these renewable energy sources into an efficient means of supplementing the home energy needs. Any homeowner that decides to pursue these options must understand that safety is an important concern. Dealing with these systems is not for the novice. However, a well thought out renewable energy plan can help supplement the home energy needs for several years.

Chapter 2
Units of Measure

Abstract *Units of measure* is the language of home energy information. We talk about our energy consumption in terms of specific units used to measure the consumption. Measuring is a vital part of the home energy information cycle. It is what guides us toward the actions we need to take in order to conserve energy and protect the environment. The ability to express those measurements in standard units has long been a benefit in understanding energy consumption. Each type of fuel such as electricity, natural gas, propane, fuel oil and others all have their own specific units associated with respect to measuring their consumption. Other environmental factors such as outside temperature affect our energy consumption. We measure these as part of the home energy information cycle and they also have specific units associated with them. We *derive* other values from the measurements to express consumption in ways that are convenient to make comparisons over time and from one location to another.

2.1 Electricity Units

Kilowatt-hours (kWh) is the standard unit of measure for electricity. It is not the easiest to understand primarily because it describes electric consumption over time. In other words, it is a *rate* of electric consumption. This is how the electric companies charge us for the electricity we use. They track the number of kilowatt-hours of electricity we use over the whole month and then charge us a standard rate (usually 10–20 cents) per kWh. We derive the unit kilowatt-hour from the unit of measure called a *watt*. Electric power expressed in watts is the rate electric energy moves through a circuit. One watt is equal to one joule per second. And 1000 W is equal to 1 kW.

The watt was named after James Watt, a British engineer and inventor. He lived from 1736 to 1819. The work of James Watt with respect to power resulted in the steam engine that powered the industrial revolution [6]. However, most of what we know today about electricity comes from Benjamin Franklin, a noted statesman and inventor who lived from 1706 to 1790. Franklin's experiments in positive and negative electrical charges resulted in the electrical lighting we enjoy today [7].

D. C. Green, *Home Energy Information,* SpringerBriefs in Energy,
DOI 10.1007/978-3-319-11349-4_2

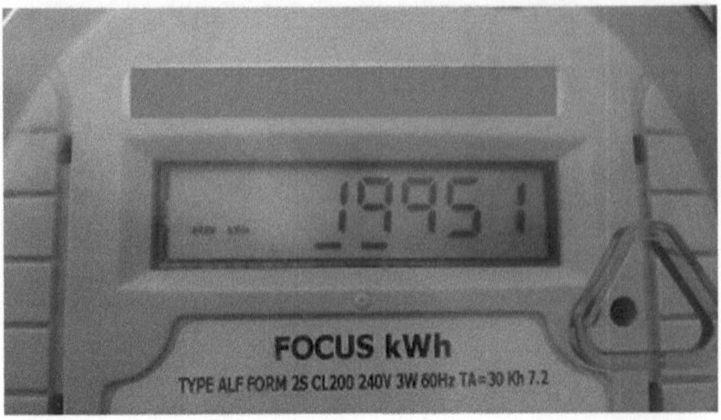

Fig. 2.1 Home electric meter

The manufacturing industry rates light bulbs in terms of watts. A 100-watt bulb is brighter than a 40-watt bulb and it also consumes more electricity. If we left a 40-watt light bulb on for 30 days continuously it would consume 28.800 kWh of electricity. My electric company charges about 10.9 cents per kWh. So, it would charge me $ 3.14 just for leaving that light bulb on.

- 30 days × 24 h = 720 h × 40 W = 28,800 watt-hours
- 28,800 watt-hours / 1000 = 28.8 kWh × $0.109 = $3.14

Newer LED light bulbs produce the same amount of light but consume much less electricity. The 40-watt light bulbs I have produce 465 lumens of light compared to the 450 lumens of light my 6.5-watt LED light bulbs produce. This could amount to a considerable savings in energy and money over time. Using the example above, if I left my LED light bulb on for 30 days continuously it would only consume 4.68 kWh of electricity and only cost me $ 0.51.

- 30 days × 24 h = 720 h × 6.5 W = 4,680 watt-hours
- 4,680 watt-hours / 1000 = 4.68 kWh × $0.109 = $0.51

One way to watch your home's electric consumption in real-time is to watch your electric meter. Most electric meters show your electric consumption in kWh. This could be a time consuming task. Depending on what is happening in your home it could take several minutes to consume even one kWh of electricity (Fig. 2.1).

There are other devices available that you can use to watch the electric consumption of small appliances in real-time. Watching your electric consumption in kWh can help you be more aware of how much energy your home uses each day.

2.2 Natural Gas Units

The therm is the standard unit of measure for natural gas. Natural gas meters actually record the consumption by volume as 100s of cubic feet (CCF). But, gas companies usually charge us by the *therm.* 100 CCF is equal to approximately one *therm*. The word "*therm*", derived from British Thermal Units (BTU), represents the potential heat generated from the fuel. One *therm* is 100,000 BTUs [8].

Converting the CCF from the meter to therms on your gas invoice is a little tricky. The exact conversion factor depends on the mix of hydrocarbons in the natural gas. This factor can vary from month to month. So, the gas company will likely show the conversion factor, called *thermal factor*, on your invoice as well. Then they calculate therms as CCF divided by the thermal factor to get therms. Here's an example of the calculation portion of a typical natural gas invoice.

CURRENT BILL ITEMIZED
In 32 days you used 57 therms
07/19/2005 reading ACTUAL 61105
06/17/2005 reading ACTUAL 61050
CCF Used for METER# xxxxxxxx 55
Thermal Factor × 1.0448
Total therms used 57

Then of course the gas company charges you a certain amount for each therm used. This amount varies by company and depends a great deal on availability of natural gas in the region where you live.

Since it is difficult to relate the readings on your gas meter to your gas bill, another great way to track your energy consumption is to use the information on your invoices. Both electric and gas companies usually include a "use history" section that shows how much you have consumed in previous months. This is how you can compare your consumption from month to month. It's important to keep in mind that billing months may have a different number of days. Table 2.1 is an example of use history in therms for a natural gas invoice.

Watching your gas consumption month to month makes it easy to spot the increased consumption during the colder winter months. This is an excellent example of how the environment effects our energy consumption. Reducing consumption amounts to a battle against the environment. If you live in an area with extreme outside temperatures, either cold or hot, you have "extreme challenges" in reducing your energy consumption.

2.3 Propane and Fuel Oil Units

Gallons are the standard units of measure for propane and fuel oil. These liquid fuels are contained in a tank until you consume them. Replenishment occurs much like the fuel in the gas tank of your car. When you get low on fuel you fill the tank up again. This likely doesn't occur on a regular basis. So the service companies will likely not send you an invoice on a regular basis like with electric and gas. How-

Table 2.1 Gas use history

Gas use history		
Bill date	Days	Therms
Jul 05	32	57
Jun 05	31	141
May 05	32	179
Apr 05	29	386
Mar 05	29	1058
Feb 05	28	1358
Jan 05	34	1174
Dec 04	29	509
Nov 04	28	380
Oct 04	30	120
Sep 04	32	56
Aug 04	31	55
Jul 04	32	69

ever, the propane and fuel oil providers charge you a set rate per gallon of fuel just like the other providers. The rate per gallon is dependent on availability of the fuel in your area.

Since your invoice arrives at varying intervals it is not much help in tracking your consumption on a monthly basis. You have to rely on the gauges attached to the tanks themselves along with the amounts (in gallons) added on each delivery.

You can still make an attempt to track your propane and fuel oil use by the month by using a systematic method of checking the gauge on your tank regularly. For instance, check the gauge at the same time of day on the first day of each month and record the gallons in the tank. You also must record the gallons in the tank just prior to each time your provider fills it along with the number of gallons added during the fill. Propane tank gauges like in Fig. 2.2 show the volume of liquid in the tank as a percentage. So you must know the total capacity of the tank to calculate the gallons available. The gauge in Fig. 2.2 attached to a 100-gallon tank has 80 gallons of propane inside. If this were a 400-gallon tank it would have 320 gallons of propane inside.

Fuel oil gauges show that the volume in the tank in gallons. Figure 2.3 shows what a typical fuel oil tank gauge might look like for a 200-gallon tank.

The Table 2.2 shows how to record these values in a log and determine your consumption in gallons for each month.

Using the Propane Use Log in Table 2.2 you can see that this home used 77 gallons in January, 56 gallons in February, 55 gallons in March and so on. Keeping track of your propane and fuel oil consumption in this manor will help you insure that your consumption is not outside the normal range. An abnormally high usage during any one month might signal a problem with your heating equipment or a leak in the tank or lines.

Fig. 2.2 Propane gauge

Fig. 2.3 Fuel oil tank gauge

2.4 British Thermal Units

The great thing about British Thermal Units (BTU) is that we can convert any of the units we talked about previously to the BTU. That is because they all involve transfer or release of heat when consumed. Scientists initially defined the BTU as the amount of heat required to raise the temperature of one pound of water one degree Fahrenheit. However, they found that it depended on the initial temperature

Table 2.2 Propane use log

Propane use log			
Date	Gallons in tank	Gallons added	Gallons used
1/1/2014	152		
2/1/2014	75		77
2/10/2014	49	120	26
3/1/2014	139		30
4/1/2014	84		55
4/15/2014	66	100	18
5/1/2014	155		11
6/1/2014	146		9

of the water. So, since 1956 its definition has been approximately equal to 1055 J or 252 g calories [9].

The BTU has become a standard for measuring energy consumption because we can convert the consumption measurement for any fuel to BTUs. This allows us to combine the consumption of different fuels and arrive at a value that represents the consumption for the whole home. The best web site I've found for converting these values is: http://www.onlineconversion.com/energy.htm [10]. On this web site you can find the following conversion factors:

- 1 KWh = 3412.1414799 BTU
- 1 therm = 100,000 BTU
- 1 gallon [U.S.] of LPG = 95,500 BTU
- 1 gallon [U.S.] of #2 fuel oil = 138,874.15823 BTU

My home uses electricity and propane. So, in order to find the total number of BTUs my home used in a particular month I would use the following calculations.

- $\text{Electric use (kWh)} \times 3,412.1414799 = \text{Electric BTU}$
- $\text{Propane use (gallons)} \times 95,500 = \text{Propane BTU}$
- $\text{Electric BTU} + \text{Propane BTU} = \text{Total BTU}$

This represents the total energy consumption for my entire home for one month. This is a giant leap forward in arriving at our home energy index (HEI). We still have environmental factors that effect our home's consumption in BTUs. My propane fireplace burns a lot in the winter time so that will obviously raise the number of BTUs consumed during those months. During the spring and fall neither my fireplace nor my heat pump runs. This lowers my consumption considerably.

2.5 Degree Days

Degree Days (DD) is an important factor in calculating our HEI. It is what helps us see how energy efficient our homes are no matter what the outside temperature. So when we compare months to months throughout the year we can get more consistent and meaningful values.

DD is a measure of how many days and degrees the Mean Daily Temperature was above or below a base temperature, usually 65° Fahrenheit (F). In our case we add those values up for the whole month. During colder months the degree days are called "heating degree days" because it requires heating to raise the temperature back up to the base temperature. During hotter months they are called "cooling degree days" since it requires cooling to lower the temperature down to the base temperature.

In warmer or cooler months when the number of degree days is high we can expect to use more energy to cool or heat our homes. During months when the outside temperature is near 65 °F then the number of degree days is low and we can expect to use less energy to cool or heat our homes. So, in order to eliminate the confusion of tracking our energy efficiency during different weather conditions we could track our energy consumed per degree day as BTU/DD. Here's an example for my home during the month of January.

- January Electric (kWh) $= 1,678 \times 3,412.1414799 = 5,725,573.2$ BTU
- January Propane (gallons) $= 52 \times 95,500 = 4,966,000$ BTU
- January Degree Days (heating) $= 1301$
- $(5,725,573.2 \text{ BTU} + 4,966,000 \text{ BTU}) / 1301 \text{ DD} = 8,217.9654 \text{ BTU} / \text{DD}$

Anyone can find the degree days for their location for any month at the BIZEE Degree Days web site (http://www.degreedays.net). Now, let's do this same calculation for my home in April when I have very little (if any) heating or cooling energy consumption.

- April Electric (kWh) $= 805 \times 3,412.1414799 = 2,746,773.8$ BTU
- April Propane (gallons) $= 0$
- April Degree Days (heating) $= 337$
- $2,746,773.8 \text{ BTU} / 337 \text{ DD} = 8,150.664 \text{ BTU} / \text{DD}$

As you can see the BTU/DD values for January and April are nearly the same even though the outside temperature where I live in January is very cold, sometimes way below 0 °F. In April, it's very mild averaging near 60° F. So, by using degree days as a factor in tracking our energy consumption we can get a consistent value to compare our energy efficiency month to month no matter how cold or hot the outside temperature is where we live.

2.6 Living Area Units

It's fairly obvious that homeowners measure the living area of their homes in *square feet* (SF). This represents the area serviced by the energy consuming appliances and equipment. The exact value may vary from utility to utility. For instance certain areas of a home may not be heated with natural gas but still have electricity for lights and appliances.

We define the living area of the home or a single room by the multiple of its length and width dimensions. We can measure homes serviced completely by all utilities by multiplying the length of the home by the width to arrive at an area value for each floor. If it is a two-story home, we multiply that result by two. This gets complicated for split-level homes and homes that do not provide utility services to all rooms. Another method is to calculate the area of each room and add them all together. The goal is to arrive at a living area value expressed in *square feet* (SF) for each utility provided to the home. So, if a home is all electric you just have one living area value. For a home serviced by both electric and natural gas there are two values, electric living area and natural gas living area. Also, since the size of our home could change, we need to record the living area(s) each month along with our energy consumption values.

We use the living area values to further refine our definition of the HEI. Just like the environment, the *size* of our home effects how much energy we will use. In general, large homes use more energy and smaller homes use less energy. So in order to compare our HEI with others to see how well we are doing we need to factor in the effects of size, similar to the way we factored in the effects of temperature using degree days. Ultimately, we want to find an energy consumption value for our home in BTU per DD SF. We multiply the DD and the SF values together. Let's add the living area of my home to the previous example for degree days to arrive at a January Electric Energy Index (EI) for my home.

- January Electric (kWh) $= 1,678 \times 3,412.1414799 = 5,725,573.2$ BTU
- January Electric Living Area (SF) = 2218 SF
- January Degree Days (DD) = 1301 DD
- January Electric EI $= 5,725,573.2 / (2218 \times 1301) = 1.9841757$

I can calculate the same EI for propane for my home and arrive at a Propane EI. If I add the Electric EI and the Propane EI together I have the Home Energy Index (HEI) for my home. This is a close approximation to the true energy efficiency of my home. It is a value I can use to measure how well I'm doing with regard to conserving energy. I can compare these values month to month to help me detect appliances and equipment that may not be as efficient as they should be. I can also compare my HEI with other homes to inspire a dialog with others on how to save energy.

Using these standard units of measure we've arrived at a value that represents our home energy efficiency. Each utility provided has its own specific units of measure. We use these units to record consumption values each month. Degree days values help normalize the effects of weather on our energy consumption values. Living area values normalize the effects of the size of the home on our energy consumption values. Together these units of measure allow us to create an Energy Index (EI) for each utility provided to the home. Adding the EI of each utility provided gives us the HEI for our home. The HEI is just one of the tools we need to continuously measure, conserve and monitor our energy consumption and participate in the Home Energy Information Cycle.

Chapter 3
Measuring Energy Consumption

Abstract Measuring energy consumption is a very important part of gathering the information we need to conserve energy, protect the environment and save money. In many cases, utility providers make measuring energy consumption by the month quite easy. They provide the values on the invoices they send us each month. In other cases, we can derive the amount of fuel we have used during a specific time period by recording deliveries and inventories on a regular basis. Along with our energy consumption values we must record environmental data that may effect our energy consumption over time and from one location to another. This is the beginning of our systematic method of gathering information needed to conserve.

3.1 Electric Consumption

Electricity providers give us an invoice with all the required information to record our electric consumption month to month. Some electric invoices are difficult to understand. Depending on the company's preferences, they may include references to demand charges, laddered charges and others. Some meter readings listed may include a conversion factor in order to arrive at the total kWh consumed. None of these items are important for our purposes. There are only a few items we need to pull from the invoice to make our data collection successful.

Most providers issue invoices shortly after the end of the month in which the services where provided. So for service in January we would look for an invoice arriving sometime in February. The exact dates the provider delivered service are most likely found near the meter readings section of the invoice. These dates may not be the first day of the month or the last day of the month. Utility providers cannot possibly read all the meters on the same day each month. So your invoice may have a billing cycle that is different than other customers of the same utility provider. In the example below of one section of an electric invoice we can see from the meter reading information the start date and the end date and we know it is referring to service primarily delivered in January, 2014.

D. C. Green, *Home Energy Information,* SpringerBriefs in Energy,
DOI 10.1007/978-3-319-11349-4_3

Table 3.1 Electric consumption

Year	Month	Electric use	Units	Start date	End date
2014	January	1180	kWh	1/1/2014	1/31/2014

Meter xxxxxxx
Jan 22, 2014 Actual Read 54322
Dec 21, 2013 Actual Read 54155

The easiest way to find the actual consumption value is to look for the "kWh" abbreviation somewhere near the bottom of the invoice. It may be close to the total charges. In this example you can see that the total kWh consumed was 1180.

- Standard Offer Svc .05100 × 1180 kWh = 60.18

You can also measure your electric consumption by reading your electric meter yourself provided your meter is showing kWh values. This is a good way to measure consumption for exactly the days you wish to measure as in the first day of the month through the last day of the month. Record these values in tabular form, preferably a spreadsheet or database. Here's an example of how to record the electric consumption values for each month after reading the meter yourself (Table 3.1).

3.2 Natural Gas Consumption

Natural gas providers also give us invoices explaining our consumption by the month. These invoices may also have some "extra" charges that make the invoice a little difficult to understand. However, we are only interested in a few items on the invoice in order to track our natural gas consumption.

Just like with the electric invoices there is usually a meter reading section that tells us the start date and end date of the billing cycle. In the example below, the provider has also placed the number of therms used conveniently next to the meter readings.

CURRENT BILL ITEMIZED
In 30 days you used 509 therms
1/16/2014 reading ACTUAL 56619
12/17/2013 reading ACTUAL 56420

So, as you can see the billing cycle for this month actually starts in the middle of the month. We might consider this to be January's bill since it ended in January. This would tempt us to read the gas meter ourselves on the first and last day of the month. However, with gas meters, this becomes a little more complicated. Since most gas meters show consumption in CCF you would have to convert that to therms using the thermal factor provided on the invoice. Keep in mind the thermal factor may be different from one month to the other. Here's how you might determine your gas consumption by reading your gas meter in CCF.

January Natural G as Readings

1/31/2014	Reading ACTUAL	56619 CCF
1/1/2014	Reading ACTUAL	56420 CCF
		199 CCF

Thermal Factor = 1.0489 (from invoice)
January Use = $199 \times 1.0489 = 208.74$ therms

Now, you can record these values in a similar manner as the electric values (Table 3.2).

As you can see, reading the utility meters ourselves gives us much more consistent start and end dates. This is the way we *must* measure the consumption for propane and fuel oil since these services are not *metered* services.

3.3 Propane and Fuel Oil Consumption

Propane and fuel oil are not metered services. The providers deliver them on an "as needed" basis. Customers typically call and ask for delivery when their fuel holding tanks get low. Therefore, there are no invoices for specific time periods to measure consumption. We must read and record the tank fluid levels ourselves in order to be wise consumers.

You can use a record similar to the Propane Use Log (see Table 2.2) for any such fuel delivered and stored in liquid form. Recording the levels at proper intervals will allow you to arrive at monthly consumption values. Be sure to record the level of the tank immediately before each new delivery. Remember that propane gauges show the volume of liquid in the tank as a percentage. Once you arrive at a monthly consumption amount you can record the values in the same manner as electric and gas consumption values (Table 3.3).

Table 3.2 Gas consumption

Year	Month	Gas use	Units	Start date	End date
2014	January	208.74	Therms	1/1/2014	1/31/2014

Table 3.3 Propane consumption

Year	Month	Propane use	Units	Start date	End date
2014	January	40	Gallons	1/1/2014	1/31/2014

Table 3.4 Fuel oil consumption

Year	Month	Fuel oil use	Units	Start date	End date
2014	January	36	Gallons	1/1/2014	1/31/2014

Fuel oil consumption is much easier to read from the gauge attached to the tank. These gauges are typically "float" gauges that show the gallons remaining in the tank. A fuel oil gauge that reads 120 has 120 gallons remaining inside the tank. After taking the readings from your fuel oil tank at regular intervals you can record them just like any other utility service consumption (Table 3.4).

3.4 Degree Days

Degree days are an important measure of how the environment effects our energy consumption. Fortunately, the BIZee Degree Days web site (http://www.degreedays.net) makes it easy for us to collect and record degree days for any location and any time period based on weather data from Weather Underground. Otherwise, we would have to calculate the degree days as follows:

- Mean Daily Temperature = (lowest daily temp + highest daily temp) / 2
- Degree Days (one day) = ABS(65° F – Mean Daily Temperature)
- Degree Days (one month) = SUM(Degree Days for each day of the month)

Using the www.degreedays.net web site we simply fill out the online form similar to the one in Fig. 3.1 and generate a file containing the degree days.

Follow these steps to get degree days for any one month:

1. Choose a weather station by entering your zip code and selecting the **Station Search** button. When the list of weather stations is populated pick the closest station with a solid bar and three stars.
2. Leave default settings for *Degree Day Type* as Heating, *Temperature Units* as Fahrenheit, *Base Temperature* as 65 ° F, *Breakdown* as Daily and *Period Covered* as Last 3 months.
3. Select the **Generate Degree Days** button.
4. After the message at the top of the page says "Your degree days are ready…" you can select the **Download Now** button.

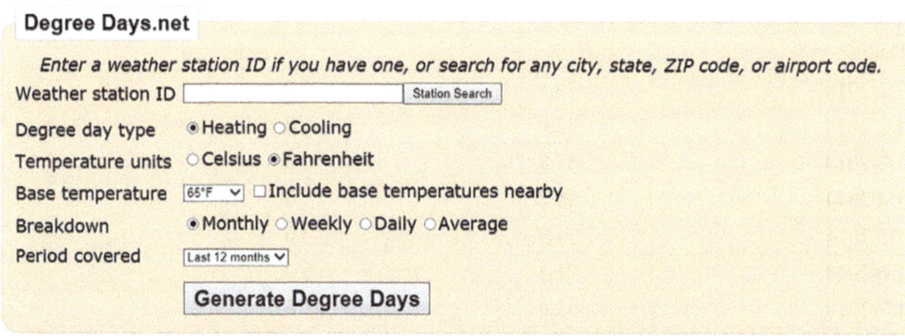

Fig. 3.1 DegreeDays.net form

5. You can save or open the.csv file with your degree days listed.
6. Total each daily value to arrive at the number of heating degree days for the entire month (Table 3.5).

You should repeat this process for *Degree Day Type*=Cooling. However, some months will not have any cooling degree days (CDD) just like some months will have no heating degree days (HDD). Add the number heating degree days to the number of cooling degree days to get the total degree days (DD) for the month.

- 1300.9 HDD + 0.0 CDD = 1300.9 DD

Now we can add degree days to our energy consumption records (Table 3.6).

3.5 Living Area

Our energy consumption is also dependent on the size of our homes. Generally speaking, a large home will consume much more energy than a small home. Of course, it is possible for a larger home that uses energy conservation measures to consume less energy than a smaller home that does not use any such conserving measures. That is why we add living area measured in square feet (SF) to our energy consumption measurements. We want to evaluate our energy consumption objectively no matter how large or small our home is.

We must measure the living area of the home separately for each utility service provided. For instance, my home has a sunroom heated by an electric heater. I also have a fireplace that runs on propane. The heat from the propane fireplace never makes it to the sunroom. So, I would calculate my *electric* living area to include the sunroom space. However, the *propane* living area would not include the sunroom space. Since heating and cooling systems consume the most energy in a home, these are the areas we consider the most when calculating living area. We should not consider a garage that is not heated or cooled but still has lights when calculating the electric living area.

Table 3.5 January heating degree days

Date	HDD	% estimated
1/1/2014	29.5	0
1/2/2014	43.6	0
1/3/2014	53.8	0
1/4/2014	35.2	0
1/5/2014	34.3	0
1/6/2014	71.3	0
1/7/2014	63.6	0
1/8/2014	44.7	0
1/9/2014	37.9	0
1/10/2014	26.1	0
1/11/2014	24.2	0
1/12/2014	25.7	0
1/13/2014	16.3	0
1/14/2014	31.4	0
1/15/2014	41.6	0
1/16/2014	38.7	0
1/17/2014	46.6	0
1/18/2014	43.5	0
1/19/2014	36.3	0
1/20/2014	27.6	0
1/21/2014	47.8	0
1/22/2014	55.3	0
1/23/2014	61.1	0
1/24/2014	55.4	0
1/25/2014	40.7	0
1/26/2014	33.6	0
1/27/2014	54.2	0
1/28/2014	61.4	0
1/29/2014	51.5	0
1/30/2014	39.4	0
1/31/2014	28.6	0
Total	1300.9	

If your entire home is uniformly heated and cooled you may be able to get your living area square footage from a tax appraisal sheet provided by your county assessor. Figure 3.2 is an example of what that tax appraisal information might look like for a two story home with a partially open cathedral ceiling.

Fig. 3.2 Tax appraisal information

	Construction	Base Area	Floor	Finished Area Sq Ft
1	Wood frame w/sh	1492	1.0	1492
1	Wood frame w/sh	761	2.0	761

Table 3.6 Energy consumption and degree days

Year	Month	Electric use	Units	Start date	End date	DD
2014	January	1180	kWh	1/1/2014	1/31/2014	1300.9
Year	*Month*	*Gas use*	*Units*	*Start date*	*End date*	*DD*
2014	January	208.74	Therms	1/1/2014	1/31/2014	1300.9
Year	*Month*	*Propane use*	*Units*	*Start date*	*End date*	*DD*
2014	January	40	Gallons	1/1/2014	1/31/2014	1300.9
Year	*Month*	*Fuel oil use*	*Units*	*Start date*	*End date*	*DD*
2014	January	36	Gallons	1/1/2014	1/31/2014	1300.9

Table 3.7 Energy consumption with degree days and living area

Year	Month	Electric use	Units	Start date	End date	DD	SF
2014	January	1180	kWh	1/1/2014	1/31/2014	1300.9	2253
Year	*Month*	*Gas use*	*Units*	*Start date*	*End date*	*DD*	*SF*
2014	January	208.74	Therms	1/1/2014	1/31/2014	1300.9	2253
Year	*Month*	*Propane use*	*Units*	*Start date*	*End date*	*DD*	*SF*
2014	January	40	Gals	1/1/2014	1/31/2014	1300.9	2253
Year	*Month*	*Fuel oil use*	*Units*	*Start date*	*End date*	*DD*	*SF*
2014	January	36	Gals	1/1/2014	1/31/2014	1300.9	2253

Table 3.8 Monthly electric consumption

Year	Month	Electric use	Units	Start date	End date	DD	SF
2014	January	1180	kWh	1/1/2014	1/31/2014	1300.9	2253
2014	February	991	kWh	2/1/2014	2/28/2014	897	2253
2014	March	823	kWh	3/1/2014	3/31/2014	629	2253
2014	April	715	kWh	4/1/2014	4/30/2014	373	2253
2014	May	474	kWh	5/1/2014	5/31/2014	271	2253

This home has a total living area of 2253 SF. Electric and propane utilities service the entire home. Once we have this information we can add living area to our energy consumption records (Table 3.7).

3.6 Recording Measurements

We want to record these measurements month after month for each utility service provided. It might be helpful to record measurements for each utility separately. A record of electric consumption for several months might look like the tabular record in Table 3.8.

Once we have all of these measurements recorded in such a manner for each utility service provided we can begin to analyze the measurements. Our analysis will turn the raw data into information used to help us conserve month after month. Our repeated measuring, analyzing and conserving are the processes that define the home energy information cycle.

Chapter 4
Analysis

Abstract Analysis is a key component of the Home Energy Information Cycle. Once we understand how to measure our energy consumption we can convert our raw consumption values into BTU units. We then combine them with degree days and living area measurements to create an energy index (EI) for each utility service. We sum the EIs into a home energy index (HEI). This is an important tool we use to measure our progress in conserving energy, protecting the environment and saving money.

4.1 Convert Consumption to BTU Units

We can convert all of the units of measure we discussed previously as British Thermal Units (BTU). This becomes the standard unit of measure for energy consumption of our whole home. Using the online conversion web site (http://www.onlineconversion.com/energy.htm), we can convert kWh to BTUs, therms to BTUs, gallons of propane to BTUs and gallons of fuel oil to BTUs. So, all of our energy consumption values will be in the same units and therefore we could sum them together if we wanted.

Let's use the factors we obtained from the Online Conversion web site in Chap. 2 to convert our energy consumption values to BTUs.

The conversion factor for electricity is:

- 1 kWh = 3,412.1414799 BTU

Our electric consumption in kWh for January 2014 was 1180 kWh. So, in order to convert that to BTUs we must multiply 1180 times 3,412.1414799.

- $1{,}180\ \text{kWh} \times 3{,}412.1414799 = 4{,}026{,}326.946282\ \text{BTU}$

The conversion factor for natural gas is:

- 1 therm = 100,000 BTU

Our natural gas consumption in therms for January 2014 was 208.74 therms. So, to convert that to BTUs we must multiply 208.74 times 100,000.

- $208.74\ \text{therms} \times 100{,}000 = 20{,}874{,}000\ \text{BTU}$

D. C. Green, *Home Energy Information,* SpringerBriefs in Energy,
DOI 10.1007/978-3-319-11349-4_4

The conversion factor for propane is:

- 1 gallon [U.S.] of LPG = 95,500 BTU

Our example of propane consumption in gallons for January 2014 was 40 gallons. So, to convert that to BTUs we must multiply 40 times 95,500.

- $40 \text{ gallons} \times 95,500 = 3,820,000 \text{ BTU}$

The conversion factor for fuel oil is:

- 1 gallon [U.S.] of #2 fuel oil = 138,874.15823 BTU

Our fuel oil consumption example in gallons for January 2014 was 36 gallons. So, to convert that to BTUs we must multiply 36 times 138,874.15823.

- $36 \text{ gallons} \times 138,874.15823 = 4,999,469.696 \text{ BTU}$

You may notice that these BTU values are very large. It would be cumbersome to use these values to compare our energy consumption from one month to another. The energy index we create will be a much smaller and more meaningful number since it normalizes the consumption with respect to temperature extremes and the size of our homes.

4.2 Calculating the Energy Indexes

The energy index describes the energy consumption for each utility service in such a way as to make it meaningful no matter how cold or hot it is or how large or small our homes. So, let's calculate the energy index for each of the utility services we now have BTU values for. Remember the formula for the energy index is:

- EI = BTU/DD SF

In Chap. 3 we found that the degree days value for January 2014 at our example location was 1300.9. In our example home the living area is the same for all utility services, 2253 ft^2. Here is how to calculate the EIs for each utility service for January 2014.

- $\text{Electric EI} = 4,026,326.946282 \text{ BTU} / (1300.9 \text{ DD} \times 2253 \text{ SF}) = 1.37$
- $\text{Gas EI} = 20,874,000 \text{ BTU} / (1300.9 \text{ DD} \times 2253 \text{ SF}) = 7.12$
- $\text{Propane EI} = 3,820,000 \text{ BTU} / (1300.9 \text{ DD} \times 2253 \text{ SF}) = 1.30$
- $\text{Fuel Oil EI} = 4,999,469.696 \text{ BTU} / (1300.9 \text{ DD} \times 2253 \text{ SF}) = 1.71$

You can see that these EI values are much smaller than the BTU values and will be much easier to compare month to month. They should vary only a small amount from month to month. These values are also very consistent from home to home.

Table 4.1 Energy index recording

Year	Month	Electric use	Units	Start date	End date	DD	SF	EI
2014	January	1180	kWh	1/1/2014	1/31/2014	1300.9	2253	1.37
Year	*Month*	*Gas Use*	*Units*	*Start date*	*End date*	*DD*	*SF*	*EI*
2014	January	208.74	Therms	1/1/2014	1/31/2014	1300.9	2253	7.12
Year	*Month*	*Propane use*	*Units*	*Start date*	*End date*	*DD*	*SF*	*EI*
2014	January	40	Gals	1/1/2014	1/31/2014	1300.9	2253	1.30
Year	*Month*	*Fuel oil use*	*Units*	*Start date*	*End date*	*DD*	*SF*	*EI*
2014	January	36	Gals	1/1/2014	1/31/2014	1300.9	2253	1.71

4.3 Recording the Energy Index

You should record the EI values for each utility service along with the other measurements for each month. In doing so, you have now recorded all the information you need to calculate your HEI except the BTU conversion factors. The BTU conversion factors are values that do not change month to month. So, they can remain stored away in the background (Table 4.1).

If you want to record these values in an electronic spreadsheet you can hide the BTU conversion factors inside a formula to calculate your EI values automatically. If the spreadsheet program automatically labels the columns of your table alphabetically from A through I and the rows numerically from 1 to 11. Then you could create the following formulas in column I to automatically calculate the EI values (Table 4.2).

- Electric EI (cell I2) = (C2 × 3412.1414799)/(G2 × H2)
- Natural Gas EI (cell I5) = (C5 × 100000)/(G5 × H5)
- Propane EI (cell I8) = (C8 × 95500)/(G8 × H8)
- Fuel Oil EI (cell I11) = (C11 × 138874.15823)/(G11 × H11)

4.4 Calculating the Home Energy Index

The home energy index (HEI) describes the energy consumption for the whole home in the same manner as the energy index does for each utility service. So, in order to arrive at the HEI we simply add the EI values for each utility service provided to the home. My home uses electricity and propane. So, if I were to calculate the HEI for my home from the example values we've seen for January 2014 it would look like this:

- 1.37 Electric EI + 1.30 Propane EI = 2.67

You should record the HEI each month along with the other measurements and EI values. You can easily sum the EI values in your table (Table 4.3).

Table 4.2 Automatic energy index recording

	A	B	C	D	E	F	G	H	I
1	*Year*	*Month*	*Electric use*	*Units*	*Start date*	*End date*	*DD*	*SF*	*EI*
2	2014	January	1180	kWh	1/1/2014	1/31/2014	1300.9	2253	1.37
3									
4	*Year*	*Month*	*Gas use*	*Units*	*Start date*	*End date*	*DD*	*SF*	*EI*
5	2014	January	208.74	Therms	1/1/2014	1/31/2014	1300.9	2253	7.12
6									
7	*Year*	*Month*	*Pro-pane use*	*Units*	*Start date*	*End date*	*DD*	*SF*	*EI*
8	2014	January	40	Gals	1/1/2014	1/31/2014	1300.9	2253	1.30
9									
10	*Year*	*Month*	*Fuel oil use*	*Units*	*Start date*	*End date*	*DD*	*SF*	*EI*
11	2014	January	36	Gals	1/1/2014	1/31/2014	1300.9	2253	1.71

Table 4.3 Home energy index recording

Year	Month	Electric use	Units	Start date	End date	DD	SF	EI
2014	January	1180	kWh	1/1/2014	1/31/2014	1300.9	2253	1.37
Year	*Month*	*Propane use*	*Units*	*Start date*	*End date*	*DD*	*SF*	*EI*
2014	January	40	Gals	1/1/2014	1/31/2014	1300.9	2253	1.30
2014	January	Home energy index (BTU/DD SF)						2.67

The HEI calculated each month tells us how well we are doing with respect to conserving energy. The goal is to make the HEI go down each month. This means that no matter how cold or hot it is we can still measure our progress. We can also compare this value (normalized for home size as well as temperature) to other homes. So, my neighbor's HEI value should be relatively close to the HEI value for my home. Of course, a little friendly competition might be helpful to inspire us both to conserve more each month.

Chapter 5
Conserving Energy

Abstract Conserving energy in the home is a methodical approach to a never ending battle with the environment. Time and weather extremes tear away at our walls, foundations, roofs and windows until they no longer hold back the effects of cold and heat. Constant vigilance is required to keep nature out and comfort in. Watching your HEI go down each month is a sure sign that your conservation efforts are paying off. Sealing the leaks, adding insulation and replacing older less efficient appliances are the big "money makers" when it comes to energy conservation. Paying attention to air flow and solar effect can breed new ideas about how to conserve. Monitor your consumption in the most minor detail to see if your conservation efforts have paid off.

5.1 Sealing Leaks

Sealing leaks in the home starts by simply walking around your home, inside and out, looking for gaps in the structure. These gaps may appear around the openings cut in the floors and walls for pipes or near baseboards down through the floor. Make sure the doors and windows seal completely when they are closed. Look for old caulking around windows and doors. When caulking gets old it will crack or shrink leaving gaps around it. Remove the old caulking and replace it. Look for missing or damaged weather stripping. These pieces of weather stripping will deteriorate over time and fall apart. If any of the weather stripping on a door or window is bad remove it all and replace it for the whole window or door. Your forced air ducts may also leak. Look for obvious gaps in the connections. Feel around electrical outlets and switch plates for any air movement. If the outlets or switch plates are leaking insulate them with an appropriate fire resistant foam sealant.

If necessary, an energy audit can pinpoint problems with leaks in your ducts and your whole house. A certified home energy audit usually lasts 4 to 6 h and costs $ 200 to $ 500. You can audit your own home by making a list of the problems as you walk around your home. You can place box fans in the windows facing out to create a draft and make it easier to feel the leaks [4].

On the outside look for cracks in the foundation and seal them with concrete filler material. I sealed the space between my foundation and siding with spray

D. C. Green, *Home Energy Information,* SpringerBriefs in Energy,
DOI 10.1007/978-3-319-11349-4_5

foam. Spray foam is a great way to seal small gaps and holes. It only takes a small amount since it expands as it cures. In the crawlspace, check for excessive dampness. You may need to add a new vapor barrier to keep moisture away from the home. Check the weather stripping around your attached garage door. You may also find that your garage door no longer fits tightly at the bottom due to your garage floor settling in spots. Try adding additional weather stripping to the bottom or laying something along the floor to seal the gap between the door and floor.

There are many types of weather stripping used to seal leaks. A rigid-jamb for windows and doors is a solid piece that you nail to the window or door frame. V-strips are adhesive backed rolls of material used to seal small gaps in sliding windows and doors. Adhesive backed foam tape comes in all sizes and shapes to fit the particular situation you need to seal. These do not last as long as other more durable types of weather stripping. Interlocking metal comes in two pieces, one attaches to the door and one to the jamb or threshold. When closed the two pieces interlock to form a seal. These may be difficult to install. You can use door sweeps to seal the area under a door that does not fit close to the threshold or floor. They attach to the bottom of the door and are susceptible to coming loose over time. Be sure to use plenty of screws when attaching them to the door [11].

Sealing the home is an important first step and you can do it without spending a large sum of money. Be diligent in your efforts and seek out every "nook and cranny" to find the gaps and holes that may be allowing air in or out of your home. Only after you have successfully sealed your home should you begin the task of insulating.

5.2 Insulate

The most important areas to insulate in the home are the attic and the crawlspace (or basement). Hot air accumulates in the attic and heats the ceiling by conduction. *Conduction* is the movement of heat through solid materials. This makes the upstairs rooms very uncomfortable in the summer time. You must insulate against that. Heat in the living area also tries to move down through the floor by conduction. Heat doesn't only rise. Insulating under the floor is a great idea as well [12].

Insulating the hot water pipes under the house will help deliver hot water to the faucets more efficiently. Ducts that travel through unconditioned spaces (crawlspaces, garages and attics) need plenty of insulation. Newer homes require R-8 insulation on ductwork but older homes may be seriously deficient in insulating value. Only after you have sealed all ductwork leaks is it appropriate to add insulation to the ducts. Duct insulation comes in three basic forms. One is the duct board which is a rigid flat piece of insulation typically used around the large rectangular return air ducts. Duct wrap comes in rolls and wraps around the ducts held in place with adhesive tape and plastic zip ties. Flexible duct insulation sleeves fit over the ducts in one piece. You must disconnect the ducts to slip the sleeve over the outside [13].

Attic insulation is especially important in hot climates. If you don't have your attic insulated properly a lot of heat builds up in the attic and makes its way down into your home. You can use fiberglass and cellulose loose-fill insulation in the attic since there is plenty of room to install 16 to 24 inches of insulation. R values vary with the type of insulation you use. But you should insulate your attic to at least a value of R-50. Safety is extremely important when handling any type of insulation. A good respirator is the safest way to insure you don't breathe in any of the fibers or cellulose material. Serious damage can occur if you breathe in these materials. There are plenty of reference materials available on how to install attic insulation. I recommend you use these references or hire a professional to do this job for you [1].

Wall insulation is something that you will definitely have to hire a professional for unless you are a contractor yourself. Older homes likely do not have the appropriate amount of insulation in the walls. However, the cost of re-insulating them is high. It is possible to blow foam insulation up into the walls without completely dismantling them. If your home sits in the middle of a large open area without any trees or shrubs to block the wind then your walls may be allowing cool or warm air to penetrate your home. Then it might be beneficial to retrofit your wall insulation. It might also be cheaper to plant some trees or shrubs to block the wind.

5.3 Ventilation

The most important concept to remember when it comes to ventilation is the *chimney effect*. At certain times of the year when it's not really hot enough for air conditioning and not cold enough for heat I open the windows upstairs and the windows downstairs. This works best in the morning and evening. The cool air rushes in the downstairs windows and the warm air rushes out the upstairs windows. Soon convection occurs like that of a chimney. The warm air rising actually *pulls* the cool air in through the windows downstairs. Sometimes I accelerate the effect by placing a fan in one of the upstairs windows facing out. Whole house fans have a similar effect by forcing air up into and out of the attic vents. The problem with whole house fans is that you have to be very careful where you put them. They may be quite noisy and they usually operate by a thermostat meaning you can't really control when they are on or off. Also, in humid environments you don't want to keep your windows open all day and draw humidity into your home, making the home more uncomfortable. During the heat of the middle of the day I close all the windows to trap the cool air in [5].

Another important type of ventilation is the ventilation in your attic. Hot air accumulates in the attic and you must provide a way for it to escape. You do this by providing air exhaust vents and air intake vents. This works similar to the chimney effect. Vents at the top of your roof let air escape and vents near the bottom let air into your attic creating a current of air out of the top vents. The vents at the top are ridge vents, wind turbines, metal roof vents or powered roof vents. Ridge vents run the entire length of your roof at the peak. Wind turbines spin in the wind to release

Fig. 5.1 Attic access panel with screen

heat faster. Metal roof vents have round or square elevated covers that allow heat to escape. Some metal roof vents have electric or solar powered fans in them to pull air out of the roof faster. The air intake vents are low under the eaves of your roof (called soffit vents) or perhaps louvered wall vents in the gable ends of your attic. It's important to keep soffit vents clean and clear of dust and debris in order to allow a good flow of air into the attic [4].

I provide a little extra air intake to the attic above my garage by adding a screen to my attic access panel as shown in Fig. 5.1. Then I leave my garage door open a few inches to allow a good flow of air up through and out of my attic above my garage. I can tell it is working by the amount of dust that collects on the bottom of the screen.

You could also install a whole house fan in place of the attic access panel. Instead of a panel with a screen you would have a fan mounted facing up into the attic. Whole house fans usually have louvers that close when the fan is not running. They also have thermostat controlled on/off switches so they only run when the temperature in the attic is hot enough that it will start to degrade the ability of your cooling system to work efficiently. That temperature is usually about 140 °F. You can also mount whole house fans on the inside (facing out) of those louvered wall vents in the gable ends of your attic. Whole house fans accelerate the chimney effect and play an important role in keeping your home *healthy*. Moving bad air out of your home and pulling fresh, clean air into your home keeps your home smelling fresh and makes for a much healthier environment.

On a smaller scale fans inside your home can help supplement the efficiency of your cooling system. Ceiling fans do a great job of moving air around your home and helping you feel cooler. This allows you to keep your thermostat set a little higher and therefore save some energy. Opinions differ on how effective ceiling fans are in the winter months. Some say if you reverse the direction of the ceiling fan in the winter it will help move warm air back down to the lower levels of your home rather than accumulate along the ceiling. I think this depends on the particular configuration of your home. I have a large cathedral ceiling in half of my home and the ceiling fan hanging from it does help move warming air down out of the upstairs and throughout the house when I use my fireplace in the winter. For the most part, fans are most useful to use while you are in the room they are located to simply cool your body temperature and make you feel more comfortable. Fans do use some electricity, so you have to weigh the cost of using these types of appliances against the benefit they provide.

5.4 Appliances

Lighting is an appliance that also requires some cost/benefit analysis. The new LED light bulbs are quite expensive, although getting cheaper all the time. One 60 W equivalent LED light bulb I purchased today cost $ 9.88. However, its packaging claims it will last 22 years! It may not last that long but it will definitely outlast the older incandescent bulbs. This 60 W equivalent bulb puts out just as much light as the older 60 W bulbs but only uses 9.9 W. So, you could use six of these LED bulbs for what it costs to use one standard 60 W incandescent bulb. These are great replacement opportunities especially for those lights that tend to be accidently left on.

One way to insure you only use appliances when needed and not unnecessarily is to use sensor switches. For instance, a motion sensor switch on a ceiling fan will insure it is only running when you are in the room and when you leave the room it turns itself off to save electricity. The same technology is available for light switches. I have several of them in my home. They make it much more convenient then manually switching the lights on and off as you leave or enter a room. This is a job that you can likely do yourself. Just shut off the circuit breaker to the room with the light switch, test the switch to make certain the power is off and replace the standard switch with a sensor switch like the one in Fig. 5.2.

Fig. 5.2 Sensor light switch

Fig. 5.3 Battery powered LED sensor light

Another way to provide light only when needed is to use battery powered sensor LED lights. These lights work great in areas that have no electric lighting or a sensor switch is not practical. My kitchen has a multi-switch panel so a sensor switch will not work there. I put a small battery powered LED sensor light on the kitchen counter that is motion activated. This works great for those quick snack trips to the kitchen in the evening. The batteries are rechargeable so the cost is minimal (Fig. 5.3).

You can automate your large power consuming appliances as well. Programmable thermostats for your heating and air conditioning system may help you save energy and money. Depending on the type of system you have, changing the temperature setting while you are away from home could save energy without sacrificing any comfort. Some people like to turn their thermostats down during the winter evenings to provide a "cozy" nights sleep. My programmable thermostat allows me to program up to four different temperature settings per day (24 hours). I program my thermostat to a higher temperature during the day in the summer time to prevent my air conditioning system from running so much. I drop the temperature setting down at night when the cool of the evening helps make it more effective. As long as I keep the house closed up in the morning the cool air remains in the house nearly all day without the air conditioning system running (Fig. 5.4).

Large appliances need proper maintenance to remain efficient. With heating and air conditioning systems it's the filters that require the most attention. Most heating and air conditioning manufacturers recommend changing or cleaning your air handler filters at least once every 3 months. Also, the inside condensing unit gives off water when the air conditioning is operating. The water must drain properly to ensure your system and your home remains healthy. A "musty" smell in the home especially around the air handling unit is a clue that there may be a problem draining the condensate water. Improper drain functioning can cause the condensing unit and blower to become dirty and therefore inefficient. If this happens it requires a trained professional to do the cleaning and repair. While the repair person is there,

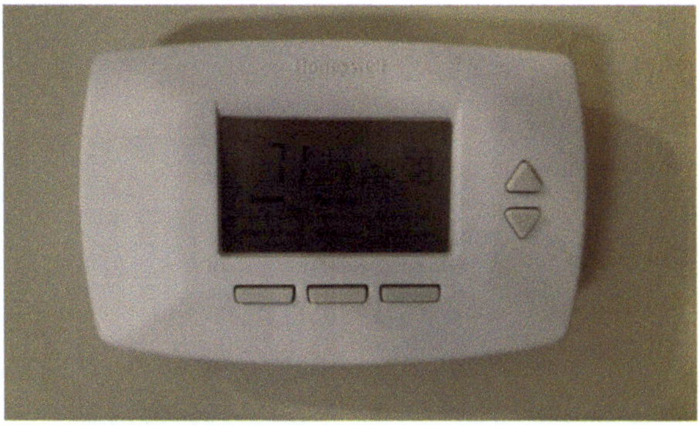

Fig. 5.4 Programmable thermostat

have them clean the outside condensing unit as well. It's important to keep it free from dust and debris, but, also not to damage the coil in the process of cleaning it.

Refrigerators are big energy guzzlers. They require the same sort of attention to the condensing coils as an air conditioning system. These coils get dusty over time and then become inefficient. Every few months pull your refrigerator out and dust off the coil on the back. Some newer models allow you to access the coil from the front. Use a vacuum cleaner to clear any dust and cob webs from under the refrigerator as well. You should never place your refrigerator close to an external heat source such as a stove. Keep the freezer full. This will stabilize the temperature and prevent the unit from cycling on and off so much. If you don't have enough food to fill the freezer then place plastic containers with water in the freezer. You can use them later when you need ice for a cooler. Don't keep an extra refrigerator in the garage. The amount you pay in electricity is more than you save buying food in bulk [14].

Clothes dryers can waste energy if you do not properly maintain them. The vents must remain completely clear for the dryer to operate efficiently. I had to replace the blower wheel on my dryer once because the vent was so blocked it stopped the wheel from turning and broke it. It seems during a snow storm snow had accumulated on top of the dryer vent (to the outside) and blocked it. The vent slowly filled with lint until it reached all the way inside to the dryer and the blower wheel. It was an expensive repair and we were having to dry our clothes twice to get them dry. Be sure to observe the dryer vent on the outside of your home regularly to ensure there is nothing blocking it.

5.5 Exterior Surroundings

You can use the exterior surroundings of your home to help conserve energy. Check with your local weather authorities to find the direction of your prevailing winds. They may be different in winter than in summer. In the winter you want to block the wind and avoid situations where objects funnel wind toward your home. Large solid objects with openings between them create the *venturi* effect and cause the

Fig. 5.5 Pergola

wind to actually speed up coming through the opening. Plants trees or park an RV between your house and the direction of the winter prevailing winds. In the summer time you want to take advantage of the *venturi* effect to create wind tunnels around objects placed in between your house and the direction of the summer prevailing winds [15]. Even small plants around your home are beneficial to cooling. They absorb heat as the air passes over them. Also, they block the reflective effects of harder surfaces that heat up quickly in the summer.

Solar heating effects are also different in winter and summer. The sun's path is further to the south in the winter. So, you want to expose your south facing windows to allow sunlight in. Expose east and west facing windows as well to absorb as much solar heating as possible in the winter. In the summer, you want to shade your east and west facing windows, especially the west facing ones to block the sun's rays. As nice as it might be to sit and watch a beautiful sunset, it may also be over heating your home. My home has its own natural shading to the west with large nut trees blocking the sun as it sets. An awning over my large west facing sunroom windows would help block the sun in late afternoon and create an even more comfortable environment. Some people build pergolas over their west facing windows and decks to create shade. Pergolas are structures, usually made from wood, that act like a roof over an area except that it is not completely covered. The pergola allows some sunlight in but blocks the majority of it. Sometimes people plant creeping vines on them to create even more cooling effect. Wisteria is a great creeping vine that covers pergolas well. Figure 5.5 shows a pergola from a great article on how to build a pergola on the Handmade Home web site (www.thehandmadehome.net) [16].

Solar screens are also an effective way to block the sun's rays and keep your home cooler in the summer and warmer in the winter. These specially made window screens block 60 to 70 % of the heat coming in through the window and actually hold heat in during the winter. They cost about $ 50.00 per window. Be sure to get screens made with a durable totally aluminum frame of at least 0.44 gauge aluminum [4]. These screens present a great opportunity to measure your energy consumption on the day before you install them and then on the day after to see if there is a difference. Of course, you want to pick two nearly identical hot summer days to perform this test.

5.6 Monitoring

Monitoring your energy consumption is an important part of conserving. It's the whole reason for the HEI. Using the HEI you can monitor your energy efficiency for your whole home each month. However, it's necessary to monitor energy consumption on a smaller scale as well. Every time you make changes to your home in a way that may effect your energy consumption you want to take detailed measurements to determine if the changes you made are actually improving your energy efficiency.

For instance, if you want to check to see how effective solar screens are then you might observe your electric meter the day before you install the screens. Pick a time of day when the sun is shining on the windows you plan to improve. Let's say from 12 to 5 pm. Make sure you are not doing other electric consuming activities during that time such as washing clothes, cooking, etc. Observe your electric meter and record the readings at 12 and 5 pm. The difference between the 12 pm value and the 5 pm value is the kWh consumed during that period. This will be primarily from air conditioning. The next morning install the solar screens and take the same readings from your electric meter after installing the screens at the same times as the previous day. Assuming the temperature on those 2 days was nearly the same, we hope too see at least some improvement in kWh consumed on the second day.

You can do the same sort of thing when you replace large appliances such as refrigerators, washing machines, dryers, ovens and microwaves. Record your electric consumption before and after making the improvements. You might want to do a similar test to see if your fireplace is actually helping or hurting your heating capabilities in the winter. If you have an electric heat pump it's easy. Just use the same procedure as described above. If you have natural gas, propane or fuel oil for heat then you might want to spread your test out over a period of a couple of weeks rather than a couple of days.

Whole house monitoring systems make it a little easier to monitor your home electric consumption for specific periods of time. These systems cost from $ 100 to $ 250 and connect either to your main electric panel or outside to your electric meter. These systems typically monitor every appliance in your home all at once. However, you can configure some of them to monitor individual circuits in your breaker box. Some systems connect to your Wi-Fi network to allow online monitoring. More recently these systems allow you to actually control your appliances from your laptop, tablet or smart phone.

Your utility company may also provide information about your consumption online. My electric company provides a nice chart of hourly consumption for any day in the recent past. Other services may provide data from your utility company for a fee and allow you to access it online. These services can create useful reports and comparisons with your neighbor's consumption.

You can test the efficiency of smaller appliances by using a plug-in monitor like the Kill a Watt, Conserve Insight or Watts Up Pro. These devices allow you to plug an appliance into them and count the kWh consumed over specific time periods [12].

Chapter 6
Monitoring Devices

Abstract Using monitoring devices to keep track of your energy consumption makes it much easier to see where energy is consumed and in what amounts. It shows us where we might start taking steps to lower our HEI. Monitoring is what helps us determine if our conservation efforts are a success. But, *monitoring devices* can also help us determine which appliances are using too much energy. We can actually prioritize our actions based on the energy used by certain appliances. It could be that our second refrigerator in the garage is really not that necessary once we find out how much energy it's consuming. Removing it could lower our HEI considerably.

6.1 Inventory

Start by taking an inventory of everything in your home that consumes energy. These appliances are primarily electric appliances but not always. You may also want to track the propane consumption of your gas grill or fireplace. And it might not always be an appliance that plugs in. You might also want to keep track of the temperature differential of your heating and air conditioning system. Basically, you should list in your inventory anything you can measure with the monitoring devices we discuss in this chapter. Here's what your inventory list might look like (Table 6.1).

6.2 Small Devices

I took these first few measurements using the *P3 Kill A Watt Electricity Load Meter and Monitor*. You can see that most of the appliances I monitored use very little electricity on their own. However, after listing all of the appliances in my home the combined consumption might be surprising. Look at the consumption of the halogen floor lamp though. It was more than my refrigerator. And, my refrigerator is old. It uses way too much electricity. Since it runs all the time, I can calculate the annual consumption of my refrigerator.

D. C. Green, *Home Energy Information*, SpringerBriefs in Energy,
DOI 10.1007/978-3-319-11349-4_6

Table 6.1 Energy consumption inventory

Item	Measurement	Units	Time period
Office TV—CRT	0.12	kWh	1 h
Computer—Notebook	0.04	kWh	90 min
Computer—LED screen	0.01	kWh	1 h
Refrigerator	0.13	kWh	1 h
Office lamp—Halogen	0.26	kWh	1 h
Floor fan	0.04	kWh	1 h
Microwave	0.05	kWh	2 min
Temp. diff.—AC	−11.0	Degrees F	6 pm
Temp. diff.—Skylight vs walls	6.0	Degrees F	Noon
Temp. Diff.—Ceiling vs walls	0.9	Degrees F	Noon
Temp. diff.—Floor vs walls	−1.1	Degrees F	Noon

$$0.13\text{ kWh} \times 24\text{ h} \times 365\text{ days} = 1{,}138.8\text{ kWh per year}$$

That is about four times what a newer, more energy efficient refrigerator would consume. Since I know my electric cost is about 10 cents per kWh, it costs me about $ 113 per year for my refrigerator. This is why keeping that old refrigerator in the garage is probably not such a good idea.

The *P3 Kill A Watt* monitoring device is a great tool for monitoring the consumption of anything that plugs into a standard 115 V AC power outlet. The device will start tracking time and kWh consumed as soon you plug it into the power outlet. It also displays volts, amps, watts, Hz and VA. This device sells for about $ 22 (Fig. 6.1).

I took the temperature differential measurements in the inventory with a *Black & Decker Thermal Leak Detector*. This is a great tool for checking the effectiveness of your heating and air conditioning system, your insulation and your windows. You can also use it to check around doors, windows, floors and ceilings for air leaks (Fig. 6.2).

You can see that the difference in temperature between my air conditioning air intake and air outlet is fairly good, more than 10 °F. It is best to take these measurements throughout your home and at different times of the day. I must have a fairly well insulated ceiling since the temperature difference between it and the walls is very small. And, the temperature difference between the ceiling and my skylight is quite large. Notice that the floor temperature is lower than the wall temp. That is the way it should be. Cool air trapped in the crawlspace overnight is pushing up against the floor. If we could only find a way to harvest that cool air and bring it up into the living area during the heat of the day.

I think if I disconnected one of my floor vents from the duct that is in the crawlspace I could open it up at certain times and let the air conditioner return air flow draw that cool air up from the crawlspace into the living area. This will only work if the outside humidity is very low. I wouldn't want to pull humid air up into the house.

Fig. 6.1 *P3 Kill A Watt* electricity load meter and monitor

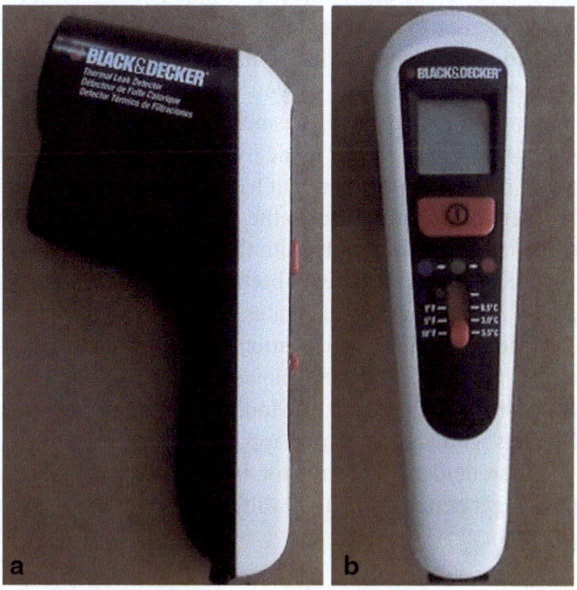

Fig. 6.2 Black & Decker thermal leak detector

This is an example of ideas you may want to try once you start monitoring your energy consumption. It is ok to experiment with ideas like this as long as you continue to monitor your consumption and try to determine if your idea is a success or failure. I can use the thermal leak detector to monitor my disconnected air vent to see if cold air is actually coming up from the crawlspace. If I had a whole house fan upstairs pushing air up into the attic and out the attic vents this idea would work even better. It would be an accelerated version of the chimney effect pulling cool air out of the crawlspace into the living area and pushing hot air out of the attic at the same time.

6.3 Whole House Devices

The most simple of all whole house monitoring devices is, of course, the meters themselves. The electric and gas meters attached to your house are great sources of information concerning your energy consumption. They can show you the consumption for any time period you wish providing you record the readings at the appropriate times. Even propane and fuel oil gauges can show you how much you have consumed for any time period.

My electric company provides information on their web site showing my consumption by the hour, day, week or month. They show the consumption in nice colorful charts along with outside temperature data to show the effect of temperature on the consumption. I'm sure some natural gas providers have the same sort of information on their web sites.

You can capture the same sort of information and provide the same reporting capabilities using any number of whole house monitoring tools. Some of them connect directly to your outside meters or to your inside circuit panel. One example is *The Energy Detective* which sells for about $ 300. You can read about it at http://theenergydetective.com. If you are technically inclined or want to monitor consumption for your small business location you might want to try the *Acquisuite Data Acquisition Server* made by Obvius. It collects data at specified intervals and stores it internally on a built in web server. So, once you connect it to a local or wireless network you can access the data from any web browser. You can also monitor temperature and humidity with the AcquiSuite.

The ThinkEco Modlet is both a monitor and controller. It plugs into a standard power outlet and communicates with your computer via a wireless USB receiver. The Modlet collects information on each outlet and allows on/off control of each outlet. Software included allows graphing of energy use over time. One receiver controls any number of Modlets around your house. You can even control the Modlets with a smart phone app [12].

These devices that monitor individual circuits are a great way to monitor the power generated from renewable energy sources such as a solar panel or wind generator.

Chapter 7
Renewable Energy Sources

Abstract Even though renewable energy sources can be expensive to install initially they can provide cheap and clean energy for many years. This makes them financially affordable in the long run. Solar power, once thought only to be advantageous in the warmest of climates is now proving to be viable almost everywhere. The same has happened to opportunities for wind power. We can now effectively take advantage of any force in nature that can generate power, store that power in batteries and supplement our energy needs. Even the chemical forces in nature allow us to create our own fuel from plant byproducts. The future is very bright for renewable energy sources in our homes.

7.1 Solar

If you intend to experiment with renewable energy the best place to start is with a small solar system. A solar panel, charge controller, golf cart batteries and an inverter can get you up and going without a great deal of cost. Once you start generating your own electricity the excitement will be overwhelming. Solar panels are available in all sizes. You can start with a small inexpensive one. Prices are about one dollar per watt. So, a 250 W panel is reasonable. You can always add more at a later time. You can do this installation yourself as long as you follow any instructions you have explicitly. It is a good idea to have a licensed professional come out and inspect your work after the system is in place.

Mount the panel on your roof if possible. It works best on a sloped roof facing south. You can also mount it on the ground. The panel has a power wire and a ground wire. The ground wire leads to a copper rod driven deep into the ground. This grounding rod protects your system from lightning. The power wire leads to your charge controller.

The charge controller connects to the battery bank and automatically adjusts the amount of power going into the batteries to prevent over charging and damaging the batteries. Charge controllers cost about $ 100. Most charge controllers will also tell you the state of the batteries, fully charged, charging or not able to accept a charge.

The battery bank can be any configuration. However, one that works well is to connect two 6-volt golf cart batteries in series (positive to negative, negative to

D. C. Green, *Home Energy Information,* SpringerBriefs in Energy,
DOI 10.1007/978-3-319-11349-4_7

positive) to make a 12 V battery bank. It's a good idea to keep an extra battery available as a spare. The cost of the batteries is about $ 400.

The power inverter connects to the battery bank via two wires with alligator clips. It converts the DC power to AC power. Power inverters come in various sizes depending on how many watts are required. Since our system has a 250 W solar panel we can only expect to get about that many watts of power out of our system at any one time. So, we add up the watts rating on all the appliances we intend to use and that tells us what size inverter to use. Let's say we'll use a 300 W inverter for this system. The inverter cost is about $ 30.

House the charge controller, battery bank, and inverter in a waterproof compartment with good ventilation on the ground level. This system will provide power to a number of small electric devices such as computers, music systems, televisions and perhaps a small refrigerator. You can upgrade the system by adding more solar panels, more batteries and a larger inverter [17].

7.2 Solar Hot Water

It's very easy to experiment with solar hot water. If you have ever left a garden hose full of water out in the hot sun for a long time and then turned on the water you noticed that the water coming out of the hose at first is very warm. The sun warms the water confined in the small space of the hose easily. You can use that same principle to supplement the heat required to heat the water in your hot water heater. However, an easier experiment is to create an outdoor shower.

To create an outdoor shower that provides warm water heated only by the sun you need only a long black garden hose (100 ft), the shower fixture, some PVC pipe and fittings. You'll need to lay the garden hose on your roof preferably on the south facing slope. I have a perfect roof on top of my sunroom that is nearly level. It only has a slight slope to it so I can just coil the black garden hose on top of the roof. For a steeper roof you might need to coil the hose around one of the vent pipes or a chimney to keep it in place. Try to avoid attaching the hose to the roof and risking creating a leaky roof. The weight of the water in the hose should hold it down.

You may want to attach two pieces of PVC pipe up the side of your house. Connect each end of the hose to each of the PVC pipes. Run one pipe to your outside faucet. You can get a "Y" splitter connection to keep your outside faucet available for other uses. Run the other pipe to your shower mounted on your back deck or maybe even to a tree. You may want to put an on/off valve in this line near the shower head if the shower fixture itself does not have one. Now you have a simple solar hot water shower.

More sophisticated systems use the same principle but use different materials. Smaller copper or black plastic tubing with aluminum reflective fins can accelerate the solar collection effects. Also, some use non-freezing solution pumped though the collection lines through a heat-transfer medium (sometimes just a water tank) to transfer the heat to the cold input line of a home water heater. In that case the cold water line coiled in the water tank collects heat from the water heated by the line running from the collector on the roof and also coiled in the water tank. These systems require a professional to install them.

Just like with solar you can expand these systems to meet the hot water needs of your home and lower your traditional energy consumption considerably. My water heater uses about 150 kWh of electricity per month. That is about $ 15 per month or $ 180 per year. So, a small investment in a solar hot water system might cut that cost in half each year. The great thing about renewable energy systems is that you can start small with little expense and use your own expertise to start saving energy and lowering your HEI.

7.3 Wind

Wind power can supplement your energy consumption from small appliances in a similar way as solar panels. The wind turns the prop which generates DC power that flows through a charge controller into a battery bank. The inverter connected to the battery bank converts the DC power to AC power for use in your appliances. So, the only question remaining is how to build the generator itself.

Wind generators must be high in the air and free from any obstructions to be effective. The higher your home is the more effective your wind generator will be. Then you need a tower to put the generator even higher off the ground. You may be able to use a short tower connected to your chimney.

You will need a DC motor that produces the highest amount of voltage at the lowest RPMs to use as the generator. A motor spinning in reverse becomes a generator. You need a motor that produces at least 1 V for every 25 RPMs. Treadmill motors seem to be popular choice. For your first experiment with wind generation use a small motor. Remember, you are mounting it high on a tower. The heavier the motor the stronger the tower needs to be.

You also need a prop with long blades to spin the motor. Making blades yourself could be a long in depth project unless you are especially good at wood working. I suggest buying a prop online for a few hundred dollars. You attach the prop to the motor.

Now you need a tail section to keep the generator facing the oncoming wind. Attach the motor to one end of a 24 inch length of square pipe using hose clamps. It's a good idea to create a motor housing out of large PVC pipe. To the other end of the square pipe attach a piece of sheet metal to act as a directional blade. Now mount the square pipe to a flange attached to a short piece of pipe large enough to fit over the tower pipe.

You can make the tower and tower base out of one inch pipe of various lengths. If you already have an old VHF antenna tower you may be able to use it instead. You hold the tower in place using guy wires attached to the ground. Connect the motor to the charge controller and you can generate free wind power. Kevin Harris has published detailed instructions on how to build this inexpensive wind generator on his web site at http://www.thekevdog.com/projects/wind_generator/ [18].

7.4 Hydro Power

Yes, you can even provide power to your home with water. Water running through a stream in an appropriate amount and with enough force can generate power. There are critical components of the system that must be present. One is the elevation change of the hill that the stream runs down. There must be considerable elevation drop or "head" to create enough force to turn the hydro turbine. You also need a lot of water during even the driest times of the year. There is a chart showing flow through nozzles of different sizes with different amounts of "head" in the Renewable Energy Design Guide and Catalog [19].

Once you've convinced yourself you have the critical components needed to generate electricity from water you'll need a considerable amount of expertise or at least an ability to follow instructions well. However, it is possible as Scott Gentleman explains in his article on the Backwoods Home Magazine web site. The article explains how he discovered the potential on his own property and took the necessary steps to make a hydro power generator. You can read the article at: http://www.backwoodshome.com/articles2/gentleman0711.html.

7.5 Storage Batteries

Batteries are extremely dangerous. You should always be careful when handling batteries of any type or size. Keep the terminals of the batteries clean and keep the connections tight. Never tamper with the connections while a battery is charging.

Almost all renewable energy sources require the use of storage batteries. These batteries save the energy produced for later use. Without that feature renewable energy would not be effective. Scientists have worked diligently over the last several years to make storage batteries more efficient by charging faster and holding the charge longer.

Lead-acid deep cycle batteries are a good choice for starter renewable energy systems because they are relatively cheap and readily available from any department store. They are the same batteries used for trolling motors, RVs and golf carts. Manufacturers design these batteries to allow them to discharge almost completely on a regular basis and not damage the battery. A car battery on the other hand is different. You must keep it nearly full of charge all the time to prevent damage to the battery.

The size of the battery bank depends on several factors including how much energy you need from it and the minimum temperature of the storage area. Temperature could have a drastic effect on the battery capacity. At 0 °F battery capacity drops as much as 50 %. Manufacturers express storage capacity in amp-hours. A 200 A h battery will supply 20 A of power for 10 h. Obviously, combining batteries will also combine the number of amp-hours and increase the storage capacity.

You can create 12, 24 or 48 V systems using 12 V batteries depending on how you wire them together. Multiple 12 V batteries wired together in parallel (positive to positive, negative to negative) will produce a 12 V system. Whereas, two 12 V batteries wired together in a series (positive to negative, negative to positive) produces a 24 V system. Then you can wire these 2-battery 24 V systems in parallel to create more capacity. In the same manner, four 12 V batteries wired together in a series (positive to negative, negative to positive) produces a 48 V system [19].

Chapter 8
Conclusion

Abstract Anyone can take steps to save energy, protect the environment and save money. It doesn't have to take a lot of time, expertise or investment to save energy. Individual homeowners are in a great position to make a difference by watching their energy consumption in detail and then taking action. As individuals we are not as limited by agreements, budgets and organizational hazards that plague big businesses that try to save energy.

8.1 Home Energy Information Cycle

The key to success in saving energy at home is the Home Energy Information Cycle. Measuring your consumption, conserving, replacing appliances and measuring again to confirm that your changes have made a difference is how you keep moving in the right direction. As long as your HEI keeps dropping you provide your own sense of inspiration to continue. You start by recording your consumption month by month. Reading your utility meters yourself on the first day of each month will give you the most accurate and easy to understand records. Seal and insulate your home as much as you possibly can. Analyze your appliances to find ones that may need replaced, repaired or removed. Add renewable energy sources to supplement your energy needs with clean and free energy. You should keep this cycle going month after month as you watch your HEI move lower. Compare your HEI to others in your neighborhood, social groups or on social media. Factors included in the HEI calculation make it possible to compare your energy efficiency to anyone in the world. Once you have mastered the Home Energy Information Cycle and watched your energy consumption drop take your experience to the workplace and share the knowledge with others.

D. C. Green, *Home Energy Information,* SpringerBriefs in Energy,
DOI 10.1007/978-3-319-11349-4_8

8.2 Home Energy Information Online

HomeEnergyInformation.com is the portal for all efforts with regard to Home Energy Information. The web site has a link to Twitter so you can follow activity in real-time by following the Twitter account @homeenergyinfo. The web site also has a link to a blog at http://HomeEnergyInformation.Wordpress.com where you can leave comments and collaborate with others on conservation efforts. Soon a link will provide secure access to a database application that will allow you to privately store your energy measurements month to month. You will be able to track your consumption, living area, weather data and even the HEI over time using a web based table. You will be able to analyze your success using colorful charts. The web site also has links to other sites that support the information contained in this eBook. Any questions, comments or suggestions are welcome. Please Tweet us, contribute to the blog or send an email to: webmaster@HomeEnergyInformation.com.

References

1. Krigger J, Dorsi C (2008) The homeowner's handbook to energy efficiency, 1st edn. Saturn Resource Management, Inc., USA
2. NWS Internet Services Team (2009) Glossary—NOAA's National Weather Service In: National Oceanic and Atmospheric Administration's National Weather Service. http://w1.weather.gov/glossary/index.php?word=mean+daily+temperature. Accessed 13 May 2014
3. Nijhuis M (2014) Can coal ever be clean? Natl Geograph 225(4):29–40
4. Cheslik D (2012) How to lower home energy bills: from A to Z, 1st edn. CreateSpace Publishing, Charleston
5. Piesse S (2011) Cool design for a comfortable home. The energy-smart house. pp 139–141. The Taunton Press, Inc. USA
6. Encyclopedia Britanica (2014) James Watt. http://www.britannica.com/EBchecked/topic/637673/James-Watt. Accessed 18 June 2014
7. Library of Congress (2010) Benjamin Franklin… In his own words. http://www.loc.gov/exhibits/treasures/franklin-scientist.html. Accessed 18 June 2014
8. American Gas Association (2014) How to measure natural gas. http://www.aga.org/Kc/aboutnaturalgas/additional/Pages/HowtoMeasureNaturalGas.aspx. Accessed 19 June 2014
9. Encyclopedia Britanica (2014) British Thermal Unit (BTU). http://www.britannica.com/EBchecked/topic/80372/British-thermal-unit-BTU. Accessed 20 June 2014
10. Online Conversion (2010) Energy conversion. http://www.onlineconversion.com/energy.htm. Accessed 1 July 2014
11. Teague M (2011) Weatherstripping. The energy-smart house. pp 85–90. The Taunton Press, Inc. USA
12. Harley B (2012) Insulate and weatherize, 1st edn. The Taunton Press, Inc., USA
13. Yates D (2011) Is your heating system an energy beast. The energy-smart house. pp 118–127. The Taunton Press, Inc. USA
14. Wilson A (2011) The energy smart kitchen. The energy-smart house, p. 167. The Taunton Press, Inc. USA
15. Numbers J (2011) Finding the sweet spot: siting a home for energy efficiency. The energy-smart house. pp 131–133. The Taunton Press, Inc. USA
16. Jamin (2013) The handmade home, how to build a pergola. http://www.thehandmadehome.net/2013/01/how-to-build-a-pergola/. Accessed 28 July 2014

D. C. Green, *Home Energy Information,* SpringerBriefs in Energy,
DOI 10.1007/978-3-319-11349-4

17. Seaman G (2012) Eartheasy, our simple DIY home solar power system. http://eartheasy.com/blog/2012/07/our-simple-diy-home-solar-power-system/. Accessed 28 July 2014
18. Harris K, /dev/thekevdog (2014) How i built a wind generator in my backyard for $ 150 http://www.thekevdog.com/projects/wind_generator/. Accessed 28 July 2014
19. Hydroelectric information, renewable energy design guide and catalog, 2006–2007. pp 45, 97. NRG Manager, USA

MIX
Papier aus verantwortungsvollen Quellen
Paper from responsible sources
FSC® C105338

If you have any concerns about our products,
you can contact us on
ProductSafety@springernature.com

In case Publisher is established outside the EU,
the EU authorized representative is:
Springer Nature Customer Service Center GmbH
Europaplatz 3, 69115 Heidelberg, Germany

Printed by Libri Plureos GmbH
in Hamburg, Germany